ENERGY TRANSITIONS
The AI–Energy Nexus

ENERGY TRANSITIONS
The AI–Energy Nexus

Lefteri H Tsoukalas

Purdue University, USA

World Scientific

NEW JERSEY · LONDON · SINGAPORE · BEIJING · SHANGHAI · HONG KONG · TAIPEI · CHENNAI · TOKYO

Published by

World Scientific Publishing Co. Pte. Ltd.

5 Toh Tuck Link, Singapore 596224

USA office: 27 Warren Street, Suite 401-402, Hackensack, NJ 07601

UK office: 57 Shelton Street, Covent Garden, London WC2H 9HE

Library of Congress Control Number: 2025033620

British Library Cataloguing-in-Publication Data
A catalogue record for this book is available from the British Library.

ENERGY TRANSITIONS
The AI–Energy Nexus

ISBN 978-981-98-2043-6 (hardcover)
ISBN 978-981-98-2044-3 (ebook for institutions)
ISBN 978-981-98-2045-0 (ebook for individuals)

For any available supplementary material, please visit
https://www.worldscientific.com/worldscibooks/10.1142/14502#t=suppl

Desk Editors: Soundararajan Raghuraman/Amanda Yun

Typeset by Stallion Press
Email: enquiries@stallionpress.com

To
Professors Mamoru Ishii, Joseph F. Pekny and Fabio H. Ribeiro
with gratitude for the dialogues on energy transitions

Preface

The global call for a "green energy transition" is often presented as an undeniable imperative: to eliminate emissions, a goal that in practice means the wholesale abandonment of fossil fuels. Unlike historical energy shifts – such as the Promethean energy transition (the mastery of fire over the past 100,000 years) or the harnessing of atomic power, both rooted in fundamental physical principles like energy density and entropy production – this contemporary transition is uniquely crafted by policy. Propelled by fears of apocalyptic climate scenarios, vast monetary incentives including subsidies and tax credits, and promises of wealth redistribution from developed to developing nations, it has garnered immense worldwide interest. This book contends that humanity's capacity to achieve this sweeping global imperative by the mid-21st century is severely limited. More critically, it argues that the pursuit of "Net Zero by 2050" could paradoxically entail dangers far exceeding the perceived risks of climate change, a concern underscored by the following points.

First, a sobering look at energy history and population reveals a stark challenge. In 1900, a world of 1.6 billion inhabitants produced negligible global emissions, largely in the absence of widespread fossil fuel consumption. To advocate for the elimination of fossil fuels by 2050 is to imply a return to an energy paradigm that historically supported a fraction of today's global population. The notion of sustaining over eight billion people on an energy footing comparable to that of 1900, without proven, hyper-scalable alternatives, raises the spectre of an unprecedented global contraction, not a sustainable transition.

Second, despite significant efforts, the empirical trend in global emissions is moving in the opposite direction of stated goals. In 2004, global CO_2 emissions stood at approximately 28.8 billion tones. Two decades later, in 2024, and after an estimated investment exceeding ten trillion dollars in the green energy transition, these emissions have climbed to 37.4 billion tones. This persistent increase, regardless of geographical shifts in emission sources, calls into question the efficacy of current strategies. The book explores the argument that addressing both global energy demand and climate risks may require a more decisive turn towards an accelerated atomic energy transition, leveraging nuclear power far beyond its currently marginal global contribution.

This book, *Energy Transitions: The Energy–AI Nexus*, offers a compelling and much-needed intervention in the ongoing global dialogue about energy futures.

Why is it Needed (Addressing Critical Gaps and Uncomfortable Truths)

1. **Challenges the Dominant Narrative:** It directly confronts the prevailing "green energy transition" narrative, arguing that its policy-driven nature and singular focus on immediate fossil fuel abandonment overlook fundamental physical, historical, and economic realities. This provides a crucial counter-perspective missing in much mainstream literature.
2. **Highlights Unforeseen Dangers:** The book dares to suggest that the relentless pursuit of "Net Zero by 2050" could introduce dangers far exceeding perceived climate change risks, forcing readers to critically re-evaluate assumptions and potential blind spots.
3. **Exposes Policy-Reality Disconnect:** It provides a sobering, data-driven assessment that questions the feasibility of current strategies given historical energy consumption patterns, population growth, and the empirical trend of increasing global emissions despite massive investments. It articulates *why* current approaches might be inadequate.

4. **Timely and Urgent:** The global energy crisis, geopolitical instability impacting energy markets, and the accelerating energy demands of AI make a rigorous, non-conventional analysis like this essential right now.

Unique Aspect of Approach and Competitive Advantage

1. **Deep Historical & Physical Grounding:** Unlike many books that focus solely on policy or economics, this work anchors its arguments in the fundamental physics of energy (energy density, entropy production) and a sweeping historical perspective of energy transitions (Promethean, Atomic). This foundational approach provides a unique, robust analytical lens.
2. **The "Energy-AI Nexus" as a Central Thesis:** It uniquely positions the burgeoning intersection of Artificial Intelligence and energy as both a primary driver of future energy demand *and* a critical, potentially revolutionary tool for managing complex energy systems. This cutting-edge focus is largely unexplored in such depth in existing literature.
3. **Contrarian, Evidence-Based Argumentation:** The book doesn't shy away from unpopular truths. By juxtaposing historical population-to-energy ratios with ambitious future targets, and by revealing the paradox of rising emissions despite trillions in green investment, it offers a distinct, evidence-backed critique.
4. **Holistic, Interdisciplinary Perspective:** It seamlessly weaves together insights from economics (Schumpeterian creative destruction), physics (thermodynamics, entropy), history, technology (AI, grid-edge, DERs), and policy, providing a truly comprehensive and integrated understanding of energy transitions.

Topicality, Value-Added Features and Why Readers Will Choose it

1. **Navigating AI's Energy Demands:** With AI's energy footprint rapidly growing, this book offers critical insights into how this demand will shape future energy systems and, importantly,

how AI itself can be leveraged to manage grid complexity and optimize energy allocation.

2. **Proposes Concrete Solutions:** It doesn't just critique; it advocates for specific, pragmatic solutions like an accelerated *Atomic Energy Transition* and the strategic deployment of AI for "rational allocation of energy" via advanced market structures and grid-edge autonomy. This offers actionable pathways.

3. **Thought-Provoking & Debate-Stimulating:** Readers seeking intellectual challenge and a deeper, more nuanced understanding of energy's future will find this book invaluable. It's designed to spark critical thinking and necessary debate among policymakers, industry leaders, academics, and informed citizens.

Goes Beyond the Green Hype

1. For those tired of simplistic narratives, this book provides a sophisticated, multi-faceted analysis that acknowledges complexities and trade-offs, offering a more realistic roadmap for global energy security and sustainability.

Key Selling Points

(1) This book is a timely perspective on the energy transition. It offers a much-needed critical analysis of today's Net Zero goals, challenging mainstream assumptions with a rigor grounded in physics, history, and economics. For library collections seeking intellectually diverse and debate-stimulating titles on climate and energy policy, this is an essential addition.

(2) The book introduces a forward-looking discussion on how Artificial Intelligence is reshaping both energy demand and system management. It's ideal for readers and researchers interested in the intersection of technology, environment, and systems thinking.

(3) This book supports coursework and research across disciplines by combining insights from engineering, environmental science, economics, and public policy. It's suitable for advanced undergraduates, graduate students, and professionals.

About the Author

Lefteri H. Tsoukalas is a professor of nuclear engineering at Purdue University. He is the principal author of the textbooks *Fuzzy and Neural Approaches in Engineering* (John Wiley & Sons, 1997) and *Fuzzy Logic: Applications in Artificial Intelligence, Big Data and Machine Learning* (McGraw Hill, 2023). At Purdue, he co-chairs the energy transition initiative of the College of Engineering called LEAPS, which stands for Leading Energy-Transition Advances and Pathways to Sustainability. His distinguished career includes serving as head of the School of Nuclear Engineering at Purdue; chairman of the United States Nuclear Engineering Department Heads Organization (NEDHO; and in various advisory and consulting positions for notable international and national organizations, including the International Atomic Energy Agency (IAEA), the OECD-Nuclear Energy Agency, the Agency for Science, Technology and Research (ASTAR) of Singapore, and the United States Department of Energy. He is a *Fellow* of the American Nuclear Society and a recipient of the Humboldt Prize, Germany's highest honor for international scientists.

Acknowledgments

This book, *Energy Transitions: The AI-Energy Nexus*, is the result of many invaluable contributions. While no list can be exhaustive, I want to express my sincere gratitude to several individuals.

Special thanks are due to Professor Seungjin Kim, Head of Purdue's School of Nuclear Engineering, and my esteemed colleagues: Professors H. Abdel-Khalik, H. Bindra, S. Chatzidakis, C.K. Choi, A. Garner, A. Hassanein, M. Ishii, X. Lou, S. Revankar, R. Taleyarkhan, and Y. Xu.

I am deeply grateful to Dr. Arvind Raman, Dean of the College of Engineering at Purdue, whose remarkable leadership and foresight established the Purdue Engineering Initiative on the Energy Transition (LEAPS). Many thanks also to Purdue President Mung Chiang for his constructive leadership.

The past and current members of the AI Systems Lab (AISL) were instrumental in contributing, reviewing, and testing material. My appreciation goes to Drs. A. Bougaev, R. Gao, Vivek Agarwal, and to S. Pantopoulou, R. Appiah, K. Prantikos, M. Pantopoulou, and U. Cotul.

I also extend special thanks to Professors D. Bargiotas, A. Chroneos, A. Daskalopulu, A. Heifetz, G. Stamoulis, and M. Vassilakopoulos for their critical input, as well as to Professor Arden L. Bement, Professor Magdi Ragheb, Professor Sangtae Kim, and Dr. Frederica Darema.

Much of the scholarly work from AISL that made its way into this book was made possible by the intellectual and material generosity of two exceptional Purdue alumni: Dr. Kostas Pantazopoulos (Class of 1998) and Mrs. Vivi Galani (Class of 1996). Their unwavering support and inspiration have been invaluable.

Finally, and most importantly, I am immensely thankful to the outstanding professionals at World Scientific, especially Ms. Amanda Yun. Her extraordinary professionalism and skill transformed my manuscript into this book.

Contents

Chapter 1

Energy Transitions in Perspective

1.1 What Is Energy Transition?

The concept of energy transition calls for a close examination of factors involved in the Tetrad of Humanity: *energy, materials, knowledge*, and *economics*. Energy underpins much of human progress over the millennia. But without knowledge of how to extract it and use it to make materials and machines to satisfy human needs in economical ways, energy is merely a concept in physics. When we look at energy transitions holistically, two stand out: the *Promethean* and the *Atomic*. The first extracts energy from molecules, the second from atoms.

For instance, the molecule of methane (CH_4) seen in Figure 1.1 has stored energy in its hydrogen–carbon (H–C) bonds, while the atom of uranium has stored energy in the nucleus. Extracting the energy stored in these two structures is emblematic of the energy transitions they represent. In the first, energy is harvested from molecular structures, while in the second from atomic structures.

It is important to contrast further the two different stores of energy. When methane H–C bonds are severed in a process called *combustion* (replacing them with oxygen–carbon bonds), they release energy in the form of fire. Methane has 1 carbon atom with 4 hydrogens around it. Much larger molecules are possible, having several carbon atoms in a molecular skeleton with "hydrogen skin" around it (Spilberberg, 2007). Most hydrocarbons are such molecules, in many cases, the remnants of past life; hence, the name "fossil fuels" and molecular derivatives ensue from these including, but not limited

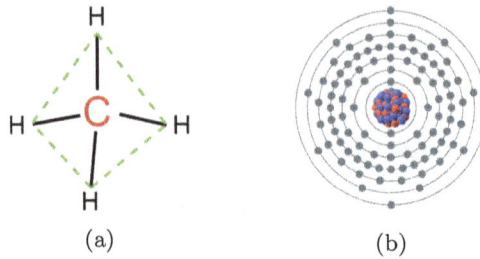

Figure 1.1.   Energy stored in (a) methane molecule and (b) uranium atom.

to, benzene (C_6H_6, with 6 carbon atoms in a ring structure and 6 hydrogens, each bonded to 1 carbon atom) and propane (C_3H_8, with 3 carbon atoms in a chain and 8 hydrogens bonded to them) (Spencer *et al.*, 2012).

In contrast, the fission of uranium releases energy in the form of heat, essentially by converting the mass of a neutron into the kinetic energy of fission products through Einstein's $E = m\,c^2$ interchangeability of energy and matter. The graphical representation of a uranium atom, as shown in Figure 1.1(b), is technically an incorrect picture of the atom. It is the archaic Bohr model of the atom that was proposed in the early 1900s before the advent of quantum mechanics (Halliday and Resnick, 1966). The actual atom is a very dynamic and complex structure modeled by quantum mechanics, while the nucleus itself is a much more complex system of many particles (235 protons and neutrons in U-235) kind of dancing with each other in modalities that are not well known and interacting with external particles such as incoming neutrons in ways that we have rather limited scientific knowledge as it belongs to the difficult "many body problem." More recent approaches, including physics-informed neural networks with transfer learning (TL-PINN) when applied to the many-body problem or other models of the nucleus, such as nuclear density functional theory (NDFT), may advance the state of knowledge of the subatomic universe (Prantikos *et al.*, 2023).

It should be noted that disassembling the nuclei of atoms produces a lot more energy per atom than severing molecular bonds. Millions of times more energy. Most importantly, breaking atoms produces not only energy but also new atoms to break, that is, it produces new fuel. And this is something new. The process of breaking atoms enables us to produce huge quantities of energy and, at the same time, breed huge quantities of fuel. This is a game-changer, a new way of obtaining energy as well as fuel for more energy in the future.

The Promethean energy transition started well over 100,000 years ago when humans learned how to make fire at will. It intensified during the last 10,000 years. The invention of agriculture gave rise to huge energy needs for machinery other than weaponry, which eventually took us to the Industrial Revolution of the 19th century. The term Promethean emphasizes the length of time it took to master fire from mythological times through agriculture and industry and on to the modern era of science. This is a pathway that correlates with the epic ascent of humanity.

Removing existential threats to humanity through fire (all predators of humans are afraid of fire), the Promethean transition slowly led to innumerable uses of hydrocarbons, building and ratcheting knowledge for agriculture to industry to solve problems of famine, disease, ignorance, and poverty. Still, there is so much more to be done with Promethean energy. A large fraction of humanity continues to live in poverty, darkness, and ignorance without the benefits of civilization, which are based on the availability of hydrocarbons.

However, the unique life-dependent centrality of hydrocarbons generates intense and seemingly contradictory fears. What will happen if we do not have enough (e.g., Malthusian fears of scarcity) or if we use them too much (e.g., fears due to having too much, a kind of cornucopia)? Although humanity without hydrocarbons will be thrown into a disastrous collapse, the Promethean energy transition is thought of by many as the *ancient regime*. This view is popular but not justified by scientific facts. As we examine in this book, the Promethean energy transition is the very foundation of Civilization. Upon it, we can build the atomic energy transition.

It should be recalled that the press release for the atomic energy transition was the bombing of Hiroshima and Nagasaki (1945), thus anchoring the atom in "shock and awe." Nonetheless, the Atomic Age that followed, while still in its infancy, holds significant promise to remove energy as a serious constraint to human growth, as it can breed fresh fuel while releasing a lot of energy.

In a fundamental sense, the only energy transition of our epoque is the nascent atomic energy transition. Yet, politics and economics have created colorful epithets for a plethora of "energy transitions." Most of them belong to the space of "communications," naming sub-transitions involving fuel substitutions, additions, or simply technology changes. For instance, in the atomic energy transition, whether we use thorium, uranium, or plutonium for fission (called "fissile materials"); or deuterium and tritium for fusion, it is not so important. The salient characteristic is not breaking molecules. We may think of thorium, uranium, or tritium fuels as additional options in the atomic energy transition. We can call them sub-transitions, or even color them as "blue" (from Cherenkov radiation) or "green" (from the renewable nature of breeding fuel), but in essence, we are only referring to the atomic energy transition.

Within an energy transition, new fuels are introduced when more energy is needed. Energy demand is the driver. For instance, building ships from steel takes much more energy than similar ships from wood. Substituting tractors for horses takes much more energy. Much more energy is used in transporting goods by plane rather than rail or ships. It takes energy and knowledge to transform material economically into useful products and services. For instance, we need material for not only buildings but also fertilizers to grow food and for medicines to cure diseases. It takes energy and knowledge to transform stones and wood into buildings and much more energy when the building material becomes steel and glass (Fay and Golomb, 2002).

For millennia, biomass such as animal dung and wood has been the main Promethean fuel. A couple of centuries ago, a steam-driven industrial revolution transitioned us from wood to coal, and a century ago, internal combustion engines drove a fuel transition from coal to petroleum, and more recently from petroleum to natural gas. The availability of energy is viewed as a significant prerequisite for the production of wealth and the dissemination of prosperity, but burning hydrocarbons (especially coal and petroleum) has produced climate concerns. Hence, a policy drive to use more emissions-free renewables, which in practical terms means hydro, solar, and wind, together with emissions-less natural gas and emissions-free nuclear. Given the episodic nature of renewables and the need for high-quality electricity to support an emerging AI infrastructure, demand for natural gas and nuclear will increase, at different speeds, to become dominant fuels as the Promethean and the atomic energy transition will overlap synergistically.

Like Promethean, the atomic transition aims to produce more energy, not less, albeit with less emissions. The overarching goal again is to support humanity's ever-growing needs for energy including, but not limited to, Artificial Intelligence (AI), the electrification of transportation, advanced manufacturing with reduced emissions, smart cities, and smart economies where electricity-fueled AI infrastructure for inferencing, advanced data centers and cloud computing, may lead to a historic transformation of the economy like the invention of Agriculture and Industry in the past (Mills, 2021).

Example 1.1. How much energy is released when breaking the hydrogen bonds in the methane molecule of Figure 1.1(a)?

Solution: Considering the methane molecule, CH_4, as shown in Figure 1.1(a), we observe that it has 4 hydrogen atoms connected to a carbon atom which is located at the center of the tetrahedral geometry of the molecule. The 4 carbon–hydrogen bonds per molecule are assumed to be of the same energy strength.

The bond dissociation energy for each H−C bond in methane is approximately (Spencer *et al.*, 2012)

$$412 \text{ kJ/mol}$$

where kJ stands for kilojoule and a mol (mole) of methane is the mass of Avogadro's number molecules of methane, i.e., $N_A = 6.022 \times 10^{23}$ molecules.

Since methane has 4 C–H bonds, the total energy required to break all the bonds in 1 mole of methane is 4×412 kJ/mol $=$ 1648 kJ/mol.

To find the energy extracted by breaking all the bonds in a single methane molecule, we use Avogadro's number:

$$\text{Energy per molecule} = \frac{1648 \text{ kJ/mol}}{6.022 \times 10^{23}} \approx 2.74 \times 10^{-18} \text{ J}$$

We can convert the energy unit from joules (J) to electron-volts (eV) by dividing by 1.602×10^{-19} J/eV. Thus, we get

$$\text{Energy per molecule} \approx 17.1 \text{ eV}$$

Hence, the maximum energy we can get from breaking all H–C bonds in a methane molecule is 17.1 eV.

1.2   Units in Energy Transition

In engineering, *energy* refers to thermal phenomena and work capacity. It is a technical concept of great importance in understanding nature and the workings of our technological civilization. Energy etymology (from Greek Energeia – Ενέργεια) points to the *action that facilitates work*. For the ancients, *pyr* (fire) was synonymous to energy, and there was a philosophical premonition about the interchangeability of energy and matter as well as the conservation of energy.[1]

In addition to fire, energy referred to forces associated with life. Medieval scholars called it the *vis viva*, which is Latin for *living force*. The modern scientific sense of energy was introduced by Thomas Young (1773–1829) in the 19th century. Its meaning is an outcome of the scientific revolutions that swept the world and gave us relativity, modern physics, and quantum mechanics (Halliday and Resnick, 1966).

Energy is measured in standard units such as Joules (1 Joule = work done by a force of 1 Newton in moving a mass of 1 kg by 1 m) or trade or legacy units such as kilowatt-hours (kWh), calories, or British Thermal Units (BTUs). For the general public, including economists, policymakers, journalists, and investors, we measure energy in monetary units, for instance, 150 dollars per kilogram of uranium. Gas, oil, coal, and transportation fuels such as benzene and kerosene are all tradeable commodities measured in monetary units. Trade or commercial units are often used in journalistic or economic discourse. The most frequent measure of energy (or fuel) is the barrel of oil, i.e., bbl or blue barrels of oil or simply barrels (1 bbl $\approx$ 159 liters $\approx$ 42 US gallons $\approx$ 0.136 tones).

For primary energy on a global scale, we often use Exajoules (EJ) and quads which stands for quadrillion BTUs (British Thermal Units). One Exajoule (EJ) equals to 10^{18} J. A Quad (Quad) is a unit of energy equal to 10^{15} BTU. It is estimated that for about 100,000 years, i.e., from the onset of the Promethean energy transition up to

---

[1]An early articulation of the interchangeability of energy and matter is found in the works of Herakleitos (535–475 BC) who speculated that all matter can be interchanged with energy and *vice versa* (in the original text "Πυρός τε ανταμοιβή τα πάντα και πύρ απάντων").

the 1850s when the industrial revolution took hold, humanity consumed a total of 13,000 quads of energy, or 13715 EJ. In the year 2023 alone, humanity consumed 620 EJ or 587 quads, while in the next 20 years, it is expected to consume more than it has consumed during the first 100,000 years of the Promethean energy transition (Judd, 2014). Hence, growing demand calls for and in a way necessitates the atomic energy transition and adjoint AI energy technologies to achieve more with less waste of energy.

The energy discourse involves mathematical abstractions that go hand in hand with material realities. It is surrounded by not only scientific standards and units but also uncertain and unknown factors and in some instances myths. It is one of the world's largest industries, and energy decisions play an important role in the calculus of economics and political considerations reaching far and wide into war and peace, prosperity and want, misery and happiness, life and death. Due to all this complexity, unlike academic and mathematical models, energy in the real world is anything but perfectly knowable.

The integration of energy with artificial intelligence, often called *Smart Energy*, aims to first, enable the emergence of an AI infrastructure and second, improve the production, allocation, and overall utilization of energy. It goes beyond physics to include inferencing, which involves data and computational approaches for optimal and resilient decisions including, but not limited to, energy, heat and thermal phenomena, mechanical work, electricity, molecular carriers, hydrogen, and energy markets (Mills, 2021).

In an abstract sense, we can think of smart energy as an AI object, a structure that centers on energy (the physical quantity) surrounded by attributes relevant to AI inferencing. The specific structures for such AI and energy, the relevant hardware, and also methods and algorithms for machine learning, discovery, and automated reasoning are an emerging, active area of innovation (Tsoukalas, 2023).

Example 1.2. What is the maximum energy in kWh that can be extracted from 1 kg of methane (CH_4)?

Solution: To calculate the energy produced from 1 gram of methane (CH_4), we need to use the heat of combustion for methane and its molecular mass. The heat of combustion of methane is the energy released when 1 mole of methane is burned completely in oxygen, forming carbon dioxide and water. The value of the heat of combustion is

$$\Delta H_{\text{combustion}} = -890.3 \text{ kJ/mol}$$

This means that 1 mole of methane releases 890.3 kJ of energy.
The molecular mass of methane is

$$\text{Molecular mass (CH}_4) = 12(\text{C}) + 4 \times 1(\text{H}) = 16 \text{ g/mol}$$

The number of moles of methane in 1 gram is

$$\text{Moles(CH}_4) = \frac{\text{Mass of 1 kg CH}_4}{\text{Molecular mass (CH}_4)}$$

$$\text{Moles(CH}_4) = \frac{1000 \text{ g}}{16 \text{ g/mol}} = 62.5 \text{ mol}$$

The energy released by 1 kg of methane is the energy released by
62.5 moles of methane, i.e.,

$$\text{Energy} = \text{Moles(CH}_4) \times \Delta H_{\text{combustion}} = 62.5 \text{ mol} \times 890.3 \text{ kJ/mol}$$

or

$$\text{Energy} \approx 55.64 \times 10^3 \text{ kJ}$$

To convert kJ to kWh, we use the conversion factor 1 kJ =
0.00027778 kWh:

$$\text{Energy} \approx (55.64 \times 10^3 \text{ kJ}) \times \left(0.00027778 \frac{\text{kWh}}{\text{kJ}}\right) \approx 15.46 \text{ kWh}$$

Hence, the maximum energy we can extract from 1 kg of methane is

$$\text{Energy} \approx 15.46 \text{ kWh}$$

The choice of units indicates that we assumed that all the molecular energy in the H–C bonds to be converted into electrical energy
through combustion (burning it). As we see in Chapter 2, this is not
the case because a thermodynamic efficiency limits the conversion
to a fraction of that (typically to about 30% in a Carnot cycle, i.e.,
we will only get approximately 5 kWh of electrical energy from a
kilogram of methane).

Example 1.3. What is the maximum energy in kWh that can be
extracted from 1 kg of uranium-235 $\left(^{235}\text{U}\right)$?

Solution: When a uranium atom undergoes nuclear fission, a massive amount of energy is released, which comes from the conversion of a small amount of mass into energy, as described by Einstein's equation: $E = mc^2$.

In the fission of U-235, a neutron splits the nucleus of U-235 into smaller nuclei and releases ~200 MeV (million electronvolts) of energy per fission event, that is to say, per atom of U-235, or more technically, per nucleus. Hence, we need to first calculate how many nuclei are there.

One mole of ^{235}U weighs 235 g and contains $N_A = 6.022 \times 10^{23}$ atoms (Avogadro's number). Hence, the energy released per kilogram of ^{235}U is the energy released by $\frac{1000 \text{ gm}}{235 \text{ gm/more}} = 4.26$ moles.

The number of atoms in 4.26 moles of U-235 is $4.26 \times 6.022 \times 10^{23} = 2.57 \times 10^{24}$ atoms. This would be the absolute number of U-235 atoms if we had a mass of 1 kg of pure U-235. Thus, the extracted maximum energy is this number times 200 MeV.

Thus, we get the following:

$$\text{Energy per kilogram} \approx (2.57 \times 10^{24} \text{ nuclei}) \times \left(200\frac{\text{MeV}}{\text{nuclei}}\right)$$

$$\approx 5.14 \times 10^{27} \text{ MeV}$$

From any elementary physics text, we obtain the conversion factor of MeV to kWh,

$$1 \text{ MeV} = 4.45 \times 10^{-20} \text{ kWh}$$

and

$$\text{Energy per kg } ^{235}\text{U} = (5.14 \times 10^{27} \text{ MeV}) \times \left(4.45 \times 10^{-20}\frac{\text{kWh}}{\text{MeV}}\right)$$

$$= 2.3 \times 10^7 \text{ kWh}$$

Therefore, the total energy release from 1 kg of U-235 is

$$\text{Energy from 1 kg of } ^{235}\text{U} = 23 \text{ million kWh}$$

This assumes that all of the energy is available. In reality, the energy available is heat; hence, we have to go through a thermodynamic cycle to convert it into motion, like the motion of electric generator turbines. Hence, only about a third, or 7.7 million kWh, will be the electricity produced.

1.3 Scarcity and Cornucopia

The prolific growth of plant life in two ancient periods of intense global warming, some 90 and 150 million years ago, left behind huge quantities of oil and gas – our principal energy supply over the last 100 years. Together with coal, this endowment of hydrocarbons, somewhat euphemistically called "fossil fuels," has been our primary source of industrial energy and raw materials or feedstock for fertilizers, pharmaceuticals, and more than 80% of all modern material.

Our chief hydrocarbon, petroleum, is relatively abundant, extraordinarily flexible, and rather easy to extract from the ground. Since the early 1900s, economic development has effectively assumed an unfailing capacity to produce, refine, and make available more petroleum in any given year than the year before. Plus-or-minus some price fluctuations on the supply side, there was always excess capacity to meet any possible growth in demand. In recent years, this assumption seems to be called into question with evidence suggesting a change in the geographical distribution of production and consumption patterns.

We have come to rely on prehistoric dead life so much that if fossil fuels were to be taken away from us, nearly 82% of all the energy that runs the world would disappear, producing, most likely, a sharp and tragic reduction in human population. In this sense, our 82% dependence on fossil fuels risks becoming a life-threatening reliance.

Since 2005, global petroleum production appears to be undergoing a phase transition where resource concentration is observed, and production becomes inelastic concerning prices (King and Murray, 2012). Some conventional oil production has plateaued or declined, rekindling earlier fears regarding production peaks and decline.

Such supply concerns coupled with global emissions targets suggest a need to decarbonize, i.e., limit fossil fuels and develop alternatives. To decarbonize economies, governments and industries are developing policies aiming at a future with less oil. For instance, the European Commission is directing members to policies that by 2050 will be practically carbon-free (European Commission, 2011). This is named "Green Transition," a policy goal to substitute hydrocarbons with solar, wind, and biomass. From a broad view on energy, this is not an energy transition. To achieve this goal, the EU and its member states have invested more than a couple of trillion euros (almost 2 trillion dollars). The results are somewhat unimpressive; hence, more recently, they have included nuclear and natural gas in the taxonomy of acceptable new fuels for energy. Again, although the term Green Transition dominates energy narratives at present, it is not based on any fundamental considerations of energy transitions – it is a policy label.

In looking at an energy transition, it is important to recall that it involves more, not less, energy, and it involves a score of consequences, some with latencies that exceed deep into the future. Smart energy is applying AI technology in the energy and material world, to extend the horizon of energy analyses and possibly improve the economics by "doing more with less." It is not a new source of energy. To address the challenges of climate change and energy security, smart energy can become an instrument for sustainability. In a world calling for more energy, fears of energy scarcity or abundance (climate security), however real or hyperbolic, call for technologies that abet the production and allocation of more energy with improved economics, a smaller environmental footprint and greater social justice, in a sustainable way.

1.4   Constraints on Promethean Energy

In the 1950s, Marion King Hubbert, while investigating possible correlations between oil exploration and production statistics, developed a method for predicting oil production peaks. His method is rather complex, and early formulations were highly mathematical as they were addressed primarily to academic audiences. In the late 1950s, however, Hubbert made a controversial prediction: US oil production would peak and enter irreversible decline in the 1970s. It turned out he was quite accurate. In 1970, US oil production peaked and has been in decline ever since. The continental 48 States of the US (excluding Alaska and Hawaii) have not been able to produce as much oil as they produced in the early 1970s until the shale revolution came along in the early part of the 21st century. Experts have applied Hubbert's theory to groups of oil fields, regions, continents, and even the world as a whole and have been pondering about the time when global oil production may peak. Some hypothesized that since global oil exploration statistics showed a discovery peak in 1965, and given that it takes roughly 30–40 years from a confirmed discovery to a production peak, we must be currently treading near the peak of global oil production (Deffeyes, 2010). Again, the shale revolution in the US and other parts of the world proved these fears were exaggerated, and in essence, they are fears of scarcity which we call Malthusian after the British economist Robert Thomas Malthus (1796–1834) who anticipated widespread famines as populations and the production of food grow at very different rates resulting in scarcity.

The feared peak in global oil production was not the point in time at which we would run out of oil but the point at which we will no longer be able to increase production and therefore maintain the system of economic growth and organization built over the last 100 years. Effects may render dysfunctional global markets that currently purport to achieve a rational allocation of energy resources using pricing signals. Sustainability is a concept implying systems consuming resources with minimal waste and thus achieving long operational lives or duration (ideally running indefinitely). In a practical sense, sustainability calls for systems and business models with elasticity margins responding to the total cost of resources.

With evidence that global oil production is increasingly more concentrated in the Middle East, Russia, and North America, concerns about energy security call for the diversification of supplies such as natural gas and nuclear and new AI technologies to do more and better with less.

Economic contraction and technological innovation are two opposing strategies to address resource scarcity. The former refers to recessions, depressions, and generally slowdowns of economies which lead to curtailed or reduced demand for energy and raw materials. David King and James Murray suggest that the debt crisis of Southern Europe is an energy crisis in disguise (2012). Greece is a case in point. Between 2000 and 2010, the Greek economy experienced unprecedented rises in energy costs (over 300% for all energy), including a 600% rise in foreign exchange allocations for oil imports. In 2000, Greece used approximately 1% of GDP for oil imports, while in 2010, oil imports took nearly 6% of GDP. Similarly, in the USA, over the first decade of the millennium, the cost of oil to the economy rose to a historical high of 8% of GDP. This created incentives for the shale revolution (Liberty Energy, Inc., 2024), where technological innovation made the US from a net importer to an exporter of energy. Similarly, leveraging atomic and AI innovations, smart energy presents an alternative for future growth with less waste and sustainable economics. It leverages AI innovation to enable the production of more primary energy and energy consumption driven by smart management of energy supply and demand, including residential needs, municipal services, commercial businesses, and industrial facilities. Most importantly, innovations enabling the atomic energy transition hold significant promise to remove energy as a constraint to human growth.

1.5 Energy Ecosystems

Energy poverty around the world is due in part to the high cost of energy. People can't afford expensive energy. Higher prices for fossil fuels are an impediment to their use. Instead of providing an incentive for the development of alternatives such as solar and wind, they may contribute to the growth of energy poverty and the destitution it entails. It is practically impossible to incentivize the atomic energy transition where energy poverty is on the rise because of more expensive Promethean energy.

In economics, money can stand as a proxy for energy. Life cycle analysis metrics, such as energy return on energy invested (EROEI), attempt to capture all energy and material quantities incurred during the lifetime of a process. This is useful for analysis. It is calculated based on average and costs (in theory, energy inputs and outputs). For planning and investment decisions, total and average costs are fine. In competitive markets, however, buyers and sellers settle with marginal, not average, costs.

Consider, for example, alternative transportation fuels and technologies. In theory, there are many. In practice, few are likely to show merit and even fewer can be economically viable. All require copious quantities of primary energy to begin with. And some require land, water, and significant resources that may be expensive and come with location-specific constraints.

Electric vehicles using rechargeable batteries, e.g., are limited by expensive minerals, such as lithium, or even less expensive ones but in high demand, such as nickel or lead. Technical means for interfacing with the power grid are available, but the grid must be capable of delivering electricity to the right place at the right time at the right price, and the electricity must be generated from a primary source such as coal, nuclear, wind, solar, or hydro and delivered through a transmission-distribution network. From an economic point of view, a series of decisions need to be made over a range of time scales including investment, sales, and cost recovery to ensure viability.

The situation is illustrated in Figure 1.2, where several pathways from carbon-free primary energy sources to different energy transformations, carriers/storage, and vehicle engines are sketched out in what is usually called an energy ecosystem.

In an energy ecosystem, biofuels may be a viable alternative to conventional transportation fuels. But cultivating crops for energy may require significant quantities of energy, water, and land, and therefore, when all is put together, EROEI may be low. Technological efforts are underway to exploit plant genetics and develop new production approaches which in combination with market-driven considerations (reflecting marginal costs) may facilitate the emergence of economically viable alternatives, but again, this may be of diminished importance in addressing energy poverty (Fay and Golomb, 2002).

In Figure 1.2, we depict alternatives and possible pathways to engineer non-oil-based transportation fuels and technologies. Which ones will prove viable and eventually acceptable will depend on economic decisions made by individual consumers, deciding not on average cost but on the price available to them at the time of a transaction. This price usually reflects the marginal cost of energy, but it may not embody several issues related to social justice, e.g., energy poverty, or environmental constraints, such as emissions targets, and it may not point to which one is more important and in what time scale.

In competitive energy markets, energy suppliers and distributors are likely price takers, not price makers. The price is the outcome of market transactions reflecting marginal costs. The instantaneous price of a barrel of oil, e.g., is based on the price of the last available (marginal) barrel sold by suppliers over a transaction interval to satisfy a given demand.

Similarly, in competitive electricity markets, the price of electricity over a transaction interval reflects the price of the last

Carbon-Free Primary Source	Energy Transformation	Energy Carrier-Storage	Vehicle Engine

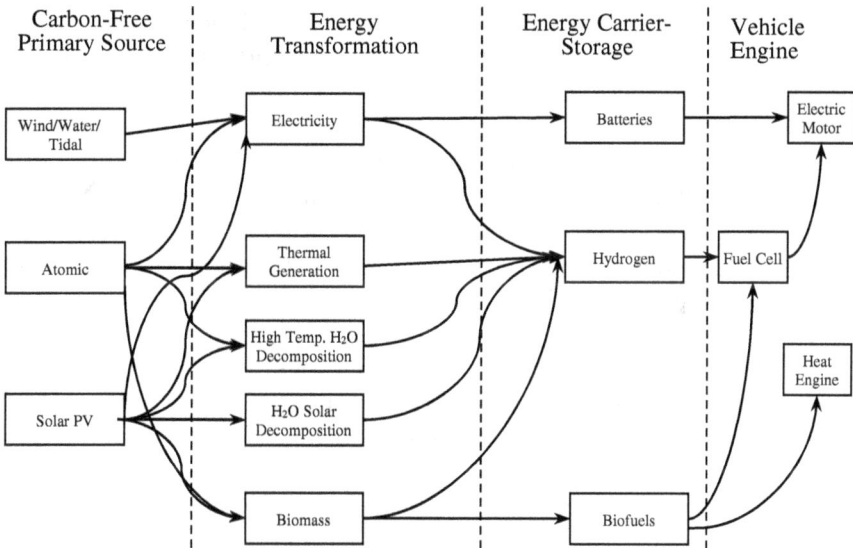

Figure 1.2. Energy ecosystem showing primary energy transformed into resources.

megawatt-hour added to the power grid through market operations that satisfy anticipated demand and at the same time maintain the electric power system in a stable and secure phase.

From a consumer point of view, smart energy adds convenience and harvests benefits for energy use endowed with scheduling flexibility. For instance, energy consumption by most appliances is price sensitive and therefore schedulable, while energy use in an operating room is price insensitive – must have it at all costs.

As a transparent example, consider an air conditioner with the ability to respond to pricing signals. The appliance is connected to the grid and the Internet. The user of the appliance has pre-programmed a set of price thresholds for electricity that can be used by the air conditioner automatically. A 24-h day-night period is divided into several fixed intervals, say 48 half-hour periods. If at a given interval the price of electricity exceeds a preprogrammed threshold of, say, 21 cents per kilowatt-hour (kwh), the air conditioner *can decide* to put itself on standby or sleeping mode until the price falls below the threshold or a critical thermostat temperature is reached. Different price thresholds may be set for other appliances. Threshold selection can be based on manual choice or be the output of machine intelligence. In the latter case, appliance use and pricing histories combined with lifestyle patterns and anticipated future variations may function as inputs with price thresholds being the outputs. The result is a higher or lower energy bill for the same comfort based on the exercise of preference and the use of technology.

Price-directed demand may include scenarios where a user is paid to use energy. For example, a plug-in hybrid automobile that can respond to pricing signals may have a negative threshold in receiving energy for battery charging. This means that the user can specify a threshold, say –3 cents per kwh, for battery charging; if the battery is offered electricity prices above 3 cents per kwh, then it will be using the energy for charging. The negative sign indicates the user is paid to consume the energy. For instance, during a low-demand period when excess wind power has kicked in unexpectedly and needs to be absorbed, a negative price threshold may indicate that the automobile's batteries are available to receive electricity and the owner will be paid to make use of the energy. Again, short-run supply–demand considerations imply that cost is determined at the margin, not on average.

Smart energy, as illustrated in the previous examples, can naturally accommodate pricing information to be used in decisions about the production, distribution, and utilization of energy. In a sense, AI-enabled energy or smart energy may achieve local solutions for using energy in globally sustainable and socially responsible ways, by not restricting energy production which may accentuate energy poverty.

1.6   Driving Principles of Energy Transitions

What drives energy transitions or sub-transitions, such as fuel additions, substitutions, and changes? Energy policy can be a driving force, but it may not be as effective, it may generate distortions, and even more important, it may have unintended consequences, the management of which may far exceed the initially intended benefits. Through policy, for instance, certain fuels can become artificially expensive or unavailable, either by restricting their supply by levying fees and tariffs on them or even by banning them, for whatever reasons. Alternatively, subsidies and tax incentives may promote other fuels and technology. Such policy tools have been suggested for addressing climate change. Yet, in practice, they seem to accentuate energy poverty, with adverse consequences for all humanity, while achieving very little in mitigating climate risks.

Policy is not likely to produce energy transitions on a sustainable matter. It is more likely to have temporary and localized effects. The policy is best exercised at national levels and is much harder to implement in multinational or even multilateral organizations like the United Nations. In the long run, policy is far less effective than fundamentals, like the driving principles outlined next located within the framework of the Tetrad of Humanity (see Section 1.1).

In addition to the principal source of energy characterizing an energy transition, molecules, or atoms, we can identify driving factors that affect and interact with the Tetrad of Humanity, such as knowledge, energy density, entropy production, demand growth, fuel availability, and supply. The goal is to have driving principles based on which we can categorize and prioritize energy transition pathways. The following five driving principles are listed as an aid to the energy transition understanding; they are by no means exhaustive.

Principle 1 – Energy Density: If the energy produced per unit mass of fuel is greater for fuel A rather than fuel B, then fuel A will prevail. In this sense, atomic fuels, such as fissile uranium, have an energy density advantage which, *ceteris paribus*, drives the atomic energy transition. We study in thermodynamic terms what the implications of energy density are and how they come into play in energy transitions.

Principle 2 – Entropy Production: Entropy production measures the amount of entropy produced during the extraction of energy. Entropy is a measure of disorder in a system, and the second law of thermodynamics states that the total entropy of an isolated system always increases over time (we discuss this in Chapter 2). This implies that any process that occurs in an isolated system will increase entropy. The manner and rate of increase can be used to uniquely identify the energy transition involved. Generally, the atomic energy transition produces less entropy than the Promethean one.

Principle 3 – Fuel Availability: This is a driving principle that rests on the availability of an energy source as well as its geographic and geological distribution. For instance, hydrocarbons are found in many places on the planet, while some nuclear fuels, such as fissile plutonium, need to be manufactured in fission reactors.

Principle 4 – Knowledge and Inferencing Support: Knowledge is used for drawing useful inferences which in turn help us build new knowledge in what is called a "ratcheting effect" in learning (Tsoukalas, 2023). Inferencing support pertains to the availability of skilled talent, both scientific and technical, and also the knowledge that the public has on energy fuels, technology, environmental impact, and the like. For instance, in the atomic energy transition, one will find widespread public distrust, apprehension, and even opposition to the technology. Hence, "public acceptance" impedes the advancement of the transition. On the other hand, AI inference support may enable humans to utilize a lot faster and in a collaborative way the knowledge that we have inherited from the past regarding scientific and empirical facts. And in turn, building successful atomic machines will grow public trust and acceptance. For instance, reactors benefit from not only the knowledge inherited from the past about nuclear fuel but also *steel* and *steam*, two materials that play a significant role in enhancing nuclear safety and efficient operations in light water reactors (LWR). AI inferencing support will help us to use such knowledge successfully and in doing so ratchet new knowledge produced for the benefit of future uses.

Principle 5 – Energy Demand: What drives an energy transition is the energy demand. With a global population likely reaching 10 billion in the 21st century, energy could be a barrier to solving the critical issues of poverty, hunger, disease, shelter, education, and safety. Hence, the call for an atomic energy transition is principally addressing the need for additional energy to address anticipated new demand.

These driving principles may be useful in developing an understanding of energy transitions anchored in the Tetrad of Humanity: *energy, materials, knowledge,* and *economics*. There are certainly more factors involved in real-world energy. However, these principles may help achieve a better understanding of energy and a more realistic prioritization of goals and objectives. For instance, what is more realistically important, achieving zero emissions by 2050 or eliminating energy poverty by 2050? These are difficult questions to answer because they involve complex scientific phenomena, economics, as well as ethical dilemmas. The prioritization of such goals based on a firm scientific analysis of facts and clarity of objectives is the practical aim of the energy transition discourse found in this book.

Problems

Problem 1.1. Although hydrogen (H) is technically not a hydrocarbon, since it is missing the carbon atom, it is often discussed as a fuel, especially in connection with "clean fuel" alternatives to fossil fuels.

(a) Estimate the combustion energy per kilogram of H and compare with the combustion energy of methane (per unit mass), propane, gasoline, and diesel. State the combustion energy in units of MJ/kg.

(b) Order the above molecules in terms of their combustion energy per unit mass and discuss what is the problem with the molecule having the highest combustion energy per unit mass.

Problem 1.2. Methane was discovered in 1776 by Italian physicist Alessandro Volta. Some decades later, stoichiometric analyses performed by chemists Antoine Lavoisier and John Dalton found that it is 74.9% carbon and 25.1% hydrogen by mass. Suppose that we have 1 kg (1,000 g) of pure methane.

(a) Use the molar mass of carbon and hydrogen to convert the grams of carbon and hydrogen into moles of each element.

(b) We know that there are always the same number of atoms in a mole of atoms of any given element. Calculate the ratio of hydrogen to carbon atoms in the sample.

(c) Using findings of (a) and (b), what is the simplest empirical formula for the molecule of methane?

Problem 1.3. Hydrogen is a diatomic molecule, i.e., it has 2 atoms of hydrogen in each molecule of hydrogen (H_2).

(a) Calculate the energy needed to dissociate the 2 atoms. This is called the dissociation energy.

(b) Given a kilogram of hydrogen, how much energy is needed to completely dissociate all hydrogen atoms in the molecules of hydrogen in 1 kg?
(c) How does this energy compare to the combustion energy produced by 1 kg of methane found in Problem 1.1?

Problem 1.4. The "hydrogen economy" envisions hydrogen (H_2) to be a carrier of energy and a "clean fuel."

(a) Why is hydrogen considered a "clean fuel?"
(b) How does the combustion energy of hydrogen compare to the dissociation energy (see Problem 1.3)?
(c) The United States consumes approximately 2 bcm (billion cubic meters) of natural gas per day. If this was to be substituted entirely by hydrogen to produce the same energy result, how much hydrogen would be needed, in bcm per day?

Problem 1.5. Nuclear fusion has two major sub-transitions in the atomic energy transition. In fission reactors, fuel material called "fissile" uses neutrons of appropriate energy to participate in a "chain reaction," which is possible because neutrons produced in the fission of the fissile fuel go on to fission new fuel which produces more neutrons, etc. In the fission of uranium-235 (^{235}U), 2.43 neutrons per fission are produced. However, they are produced with kinetic energies of approximately 2 MeV; hence, they are called "fast neutrons." In order to fission new ^{235}U nuclei, their energy needs to be moderated, i.e., reduced to a "thermal" range which is approximately 0.025 eV. Hence, neutrons with such kinetic energy are called "thermal neutrons." In a reactor, moderation takes place by neutrons losing their energy through scattering with a moderator, typically water or graphite.

(a) If in any scattering a fission neutron loses 60% of its energy (on average), how many scattering interactions does it need to make to reach 0.025 eV?
(b) The scattering events in (a) are considered highly inelastic. Why?
(c) What would be an elastic collision (interaction) between a fast neutron and a hydrogen atom?
(d) If the most probable speed for 0.025 eV neutrons is 2200 m/s, how would you calculate the mass of a neutron?

Problem 1.6. What is the maximum energy in kWh that can be extracted from 1 kg of uranium-233 (233U)? Assume that the energy per atom fissioned is similar to the energy in Example 1.3.

Problem 1.7. The fissile material in the Hiroshima bomb (code-named "Little Boy") was about 6 kg of plutonium-239 (^{239}Pu).

(a) Calculate how many atoms are there and assume that the energy produced per atom in the nuclear fission of ^{239}Pu is 200 MeV.
(b) Calculate the maximum energy in *kilotons of TNT equivalent* that can be extracted from 6 kg of plutonium-239, using the conversion factor, 1 kiloton of TNT = 4.184×10^{12} J.
(c) The actual energy produced by "Little Boy," called the "yield" of the bomb, was estimated to be 15 kilotons of TNT equivalent. Is this different from your answer above, and if so, how do you explain this difference?

References

Deffeyes, K.S., *When Oil Peaked*, Hill and Wang, New York, NY, 2010.

European Commission. Energy Roadmap 2050 (COM(2011)885final). Publications Office of the European Union. 15 December 2011. https://eur-lex.europa.eu/LexUriServ/LexUriServ.do?uri=COM:2011:0885:FIN:EN:PDF.

Fay, J.A. and Golomb, D.S., *Energy and the Environment*, Oxford University Press, New York, NY, 2002.

Halliday, D. and Resnick, R., *Physics Parts I and II*, John Wiley & Sons, New York, NY, 1966.

Judd, A.M., *The Engineering of Fast Nuclear Reactors*, Cambridge University Press, Cambridge, UK, 2014.

King, D. and Murray, J., "Oil's tipping point has passed," *Nature*, 481, 433–435, 2012.

Liberty Energy, Inc. Bettering Human Lives 2024 (3rd ed., issued January) 2024. https://libertyenergy.com/wp-content/uploads/2024/02/Bettering-Human-Lives-2024-Web-Liberty-Energy.pdf.

Mills, M.P., *The Cloud Revolution: How the Convergence of New Technologies Will Unleash the Next Economic Boom and a Roaring 2020s*, Encounter Books, New York, NY, 2021.

Prantikos, K., Chatzidakis, S., Tsoukalas, L.H., and Heifetz, A., "Physics-informed neural network with transfer learning (TL-PINN) based on domain similarity measure for prediction of nuclear reactor transients," *Nature Scientific Reports*, 13, 16840, 2023.

Spencer, J.N., Bodner, G.M., and Richard, L.H., *Chemistry Structure and Dynamics*, 5th edn., John Wiley & Sons, New York, NY, 2012.

Spilberberg, M.S., *Principles of General Chemistry*, McGraw Hill, New York, NY, 2007.

Tsoukalas, L.H., *Fuzzy Logic: Applications in Artificial Intelligence, Big Data, and Machine Learning*, McGraw Hill, New York, NY, 2023.

Chapter 2

The Promethean Energy Transition

2.1 Energy from Molecules, Heat, and the Capacity for Work

The Promethean energy transition is the epic rise of Humanity from the Animal Kingdom to Civilization. It started well over a hundred thousand years ago when early humans managed to produce *fire-at-will* gradually transforming themselves from hominids roaming in Nature to *Homo Fabiens*. Using a variety of fuels, almost all hydrocarbons, this epic transition continues to this day. *Homo Fabiens* invented agriculture and machines and materials to advance humanity's numbers, knowledge, and wealth, all based on molecules with hydrogen-carbon bonds (H–C). Wood, coal, oil, and natural gas all belong to this group. When they burn in a process called *combustion*, they undergo exothermic oxidation reactions where H–C bonds break to form carbon-oxygen (C–O) bonds and in the process release a lot of energy but also byproduct molecules like CO_2.

Energy may be defined as the capability of matter or a system to cause changes in the same or other systems. For most of human history, it was a concept cast in *anthropocentric terms*. Then, in the 1840's a breakthrough occurred: two German physicians, Robert Julius Mayer (1814–1878) and Hermann Helmholtz (1821–1894), and a British experimentalist, James Prescott Joule (1818–1889) discovered the equivalence between heat and mechanical work (1 cal = 4.18 J = 4.18 kg $- m^2/s^2$).

Very few discoveries have done as much to advance human comfort and well-being as this one. The original discovery was made to explain physiological observations such as the reduced color difference between arterial and venal blood in the Tropics[1] and heat-carrying working fluids. Energy quickly became an object of scientific interest and technical investigations followed suit, leading to the new *Science of Thermodynamics*. Numerous technical advancements in a multitude of different types of engines including the Otto and Diesel internal combustion engines revolutionized transportation. One of Thermodynamics' early achievements was to dispel the possibility of *Perpetuum mobile* (perpetual motion machines), a fascinating obsession of fraudsters. It became clear, during this period, that energy was destined to become a central theme in the narrative of the modern world, informing not only the discourse of science and technology but shaping crucial conversations in politics, economics, international relations, and geostrategic considerations.

From the 1870s to the early 1900s the work of several great thinkers such as Rudolf Clausius (1822–1888), Lord Kelvin (1824–1907), James Maxwell (1831–1879), and Ludwig Boltzmann (1844–1906), to mention but a few, made energy formulations abstract, in fact, highly mathematical and axiomatic. This accomplishment allowed energy theoretical considerations to survive the scientific revolutions of relativity, modern physics, and quantum mechanics without any major change to original theories. Even though the underlying concepts of physics radically changed, highly generic and austere mathematical formulations such as those of Constantine Caratheodori (1873–1950) allowed energy-related concepts and thermodynamics to maintain a universal mathematical description.

More recently, computer technologies have turned the energy pendulum once again towards the *anthropocentric side*, aligning energy with Artificial Intelligence (AI) in the Tetrad of Humanity, *energy, materials, knowledge*, and *economics*. Hence, the notion of energy transition has brought to the fore, lively discussions about energy availability, costs, security, and risks of different fuels. The basic message emerging from these narratives is the need for

---

[1]See *History of Thermodynamics* for a fascinating account of Robert Mayer's discovery while on a trip to Java as a ship's physician.

more energy; how to grow further the energy supply to meet an ever-growing demand at a global scale. The color adjectives that seem to proliferate as properties of energy transition, e.g., green energy transition, blue, or even gray, are communication attempts to draw attention for more energy. But there is no energy transition in a fundamental sense, they are part of an ongoing Promethean energy transition, albeit with many new aspects such as markets, supply and demand balances, and environmental risks—all based on knowledge from the Thermodynamics revolution unleashed a couple of hundred years ago.

2.2 Energy in Physics

The total energy of a system consists of the sum of kinetic, potential, and internal energy, whereas transfers of energy from one system to another consist of work and heat. Kinetic energy is directly related to motion, and the potential energy pertains to position, arrangement, and configuration. Internal energy is related to the temperature, pressure, and chemical composition of an energy carrier. The quality of energy transfer in the form of heat depends on the temperature under which the transfer occurs. Work represents the energy form with the highest thermodynamic quality because it can be converted entirely, in an ideal process, into any other energy form.

All energy forms can be converted into work. Potential energy and kinetic energy can be converted entirely into work, whereas the capability of internal energy and heat to be converted into work depends on their thermodynamic parameters such as temperature, pressure, and chemical composition.

The unit of work is the work done by a unit force in moving a body of unit mass a unit of distance in the direction of the force. In the metric system, this means a force of 1 N multiplied by a distance of 1 m that defines 1 J which is the unit of work and energy in general (1 J = 1 N–m or J = N m).

The work W done by a force $\boldsymbol{F}$ to move a body a distance $d\boldsymbol{r}$ is defined as

$$W = \boldsymbol{F} \cdot d\boldsymbol{r} \tag{2.1}$$

where $\boldsymbol{F}$ and $d\boldsymbol{r}$ are vector quantities and W is a scalar (vector quantities in this book are in bold typeface, scalars in italics).

The total work done in moving a body from an initial position r_1 to a final position r_2 is

$$W = \int_{r_1}^{r_2} \boldsymbol{F} \cdot d\boldsymbol{r} \qquad (2.2)$$

Potential energy is energy due to configuration. Consider a body of mass m and hence weight mg (where g is the acceleration due to gravity $g = 9.81$ m/s^2) which is held a distance h above the ground. Its potential energy E_{pot} which is due to this configuration is mathematically given by

$$E_{\text{pot}} = \text{mg} \cdot h \qquad (2.3)$$

If the mass m was to move with speed v, its kinetic energy E_{kin} would be

$$E_{\text{kin}} = \frac{1}{2} m \cdot v^2 \qquad (2.4)$$

For m held above ground, the release means falling a distance of h. From the physics of freely falling bodies (Halliday and Resnick, 1966), we have that its speed is

$$v^2 = 2gh \qquad (2.5)$$

and from Newton's second law, we have that $F = m\,g$ which gives $m = F/g$.

Thus, the kinetic energy E_{kin} may be written as

$$E_{\text{kin}} = \frac{1}{2} m\, v^2 = \frac{1}{2} \left(\frac{F}{g} \right) (2gh) = Fh \qquad (2.6)$$

Some important terms involved in the energy discourse are discussed next.

Power: The unit of power is the Watt (W), the rate at which a unit of energy is transferred per unit of time; that is 1 W = (1 J)/(1 s), or, 1 W is simply 1 J/s. An equivalent definition in electrical systems is given in terms of electrical or circuit current and voltage measured in Amperes (A) and Volts (V), respectively that is

1 W $= (1$ A$) \times (1$ V$) = 1$ A $\times 1$ V. In mechanical systems, horse-power is encountered as a unit of power where 1 horsepower $=$ 550 ft $-$ lb per second $\cong 0.75$ kW. In addition to Joule, we use the units of kWh, and foot-pound as units of energy. A kilowatt-hour (kWh) is the energy supplied.

Energy: It should be recalled that, according to the First Law of Thermodynamics, energy cannot be produced or destroyed. It can only be converted from one form to another. A specific form of energy, e.g., electricity can produce work of 1 kWh when an electric power source of 1 kW (1,000 Watts) operates for 1 hour.

Energy Equivalence: Quite often, we speak in practice of work or another energy form in terms of the fuel required to produce an equivalent amount of work. For example, one barrel of oil, if burned completely, would release 1.7 MWh (megawatt-hours). Hence, we say that one barrel of oil equivalent (boe) produces 1.7 MWh or 1 boe $= 1.7$ MWh. Similarly, 1.5 tons of coal is equivalent in an energy sense to approximately 12,000 kWh.

Force: The unit of force is the *Newton* (N) which is the force required to accelerate a mass of 1 kg by 1 m/s^2.

Entropy: This is a thermodynamic measure of disorder or random-ness in a system and stands as a measure of the system's energy that is unavailable for work. In simpler terms, it's a measure of how spread-out energy is and how difficult it is to harness. For instance, in a highly ordered system, like a crystal, the particles are arranged in a specific pattern. In a disordered system, like a gas, the particles are randomly distributed.

Entropy Production: This is the rate at which entropy increases and stands as a measure of the irreversibility of a process. Irreversibil-ity arises from factors like friction, heat transfer across a temperature difference, and chemical reactions. These processes dissipate energy, making it less useful. Entropy production is used in studying sus-tainability. The cumulative entropy production in a closed system is a measure of how much useful energy is left in the system. In bio-logical systems, entropy production can indicate how far along the aging process has advanced. Both entropy and entropy production will be examined in this section as measures of energy transitions and sub-transitions.

Consider next a simple example that illustrates the conversion of kinetic energy into potential energy and vice versa.

Example 2.1. A ball of mass $m = 1$ kg falls from a distance $h = 1$ m above the ground as shown in Figure 2.1. At the extreme points of its path, that is the lowest point (the ground) and the highest point (the shelf) the total energy of the ball is entirely potential energy (due to its position) and kinetic energy, respectively. What happens in between and what is the speed with which it hits the ground?

Solution: At intermediate points, the energy (potential and kinetic) it is a mixture of both. Total energy remains constant because of the principle of conservation of energy which is also known as the *First Law of Thermodynamics:* Energy changes (is converted) from one form to another but the total energy remains constant.

At the top of the shelf, the energy is entirely potential and is given by Eq. (2.3), that is,

$$E_{pot} = mg \cdot h = (1 \text{ kg}) \left(9.81 \ \frac{m}{s^2} \right) (1 \text{ m}) = 9.81 \text{ J}$$

Due to conservation of energy and assuming no energy losses due to air resistance, the potential energy E_{pot} at the top is converted entirely into kinetic energy E_{kin} at the ground, which is given by

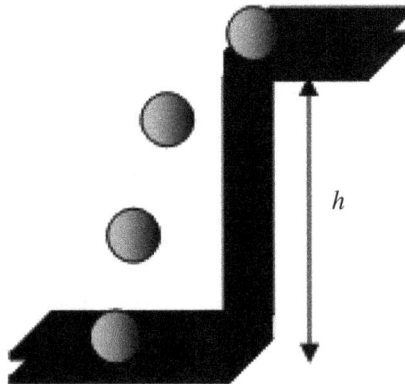

Figure 2.1. A ball falling a distance h provides a simple example of the continued conversion of potential energy into kinetic energy.

Eq. (2.4) that is,

$$E_{kin} = \frac{1}{2}m \cdot v^2$$

hence, $E_{pot} = E_{kin} = \frac{1}{2}m \cdot v^2 = 9.81$ J, and we can calculate the speed,

$$v = \sqrt{2\frac{E_{pot}}{m}} = \sqrt{2\frac{9.81 \text{ j}}{1 \text{ kg}}} = \sqrt{2\frac{9.81 \text{ kg}\frac{m^2}{s^2}}{1 \text{ kg}}} = 4.43\frac{m}{s}$$

At the top of the self, the ball has only potential energy, hence zero speed, as it falls its speed increases by 9.81 m/s every second (due to gravity which accelerates falling objects with an acceleration of $g = 9.81 \frac{m}{s^2}$).

Finally, when the ball hits the ground, all energy is kinetic and the impact speed is $v = 4.43 \frac{m}{s}$.

2.3   Energy in Thermodynamics

Up to this point, we have emphasized the notions of work, kinetic energy, and potential energy. These are notions quantified and formulated mathematically in the science of mechanics. In thermodynamics though, and in energy conversion in general, we deal primarily with internal energy, U, which we examine next.

A material system can be thought of as a collection of particles (e.g., molecules but similarly for atoms) each of which behaves like a point moving in different directions and with all sorts of speeds leading to an understanding of the internal energy U of a system.

Consider a small finite quantity of energy δQ added to a system, which performs an infinitesimally small amount of work δW and in doing so, it changes the internal energy of the system by an infinitesimally small amount dU. The First Law of Thermodynamics states that the total energy associated with this process is conserved, a statement that can be mathematically expressed by the following equation (assuming that the changes in kinetic and potential energies are negligible)

$$dU = \delta Q - \delta W \tag{2.7}$$

Equation (2.7) says that the change in internal energy of a system equals the amount of heat added minus the amount of work done by the system.

One of the great stories of scientific and engineering genius is found in a little book published in 1824 by Sadi Carnot (1796–1832), a young French military engineer who studied heat and motive power. The cover page of the book, which in English translation reads "Reflections on the Motive Power of Fire, and on Machines Fitted to Develop that Power" can be seen in Figure 2.2. Carnot described an idealized engine cycle (the so-called Carnot cycle) and the corresponding maximum theoretical efficiency, without even knowing at that time the correct physical concepts of internal energy and heat. The year 1824 marks the beginning of the new science of Thermodynamics.

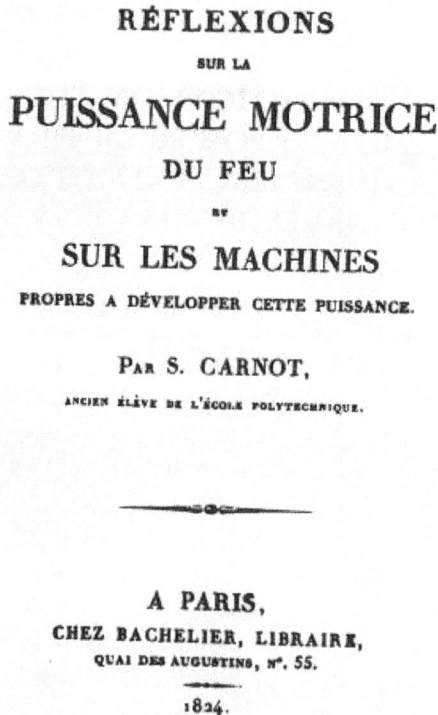

RÉFLEXIONS

SUR LA

PUISSANCE MOTRICE

DU FEU

ET

SUR LES MACHINES

PROPRES A DÉVELOPPER CETTE PUISSANCE.

Par S. CARNOT,

ANCIEN ÉLÈVE DE L'ÉCOLE POLYTECHNIQUE.

A PARIS,

CHEZ BACHELIER, LIBRAIRE,

QUAI DES AUGUSTINS, n°. 55.

1824.

Figure 2.2.   Cover of Sadi Carnot's 1824 book *Reflections on the Motive Power of Fire and on Machines Fitted to Develop that Power.*

Carnot's physical conceptualization of heat implied that it was a fluid called "caloric," and its movements from hot to cold were somewhat similar to water or objects "falling," for example, as a result of gravity similar to the ball's falling in Figure 2.1. For Carnot, the analog of elevation h is the temperature difference. He declared *wherever there exists a difference of temperature, motive power can be obtained.*

As anyone shaking or stirring a liquid easily discovers, obtaining heat from mechanical work is a straightforward process. The converse, that is mechanical work from heat, is beset with difficulties. The *Second Law of Thermodynamics* as articulated by Rudolf Clausius (1822–1888) states that *There can be no engine that can completely and continuously transform all the heat supplied to it into mechanical work.* The Second Law posits an intrinsic property called *entropy* indicated by the symbol S. The entropy of a system changes through the transport of heat and material and irreversibilities. A change in entropy of a system dS occurs when a quantity of heat δQ is supplied to it at temperature T:

$$dS = \frac{\delta Q}{T} \tag{2.8}$$

All natural and technical processes are irreversible and associated with entropy generation. The entropy generation in a system is always positive, always increasing; it approaches zero only if a process approaches reversibility. Entropy's practical implications tell us that there is an energy tax imposed by nature (the Second Law of Thermodynamics) to be paid every time a process occurs, for example, when heat is transformed into mechanical work and therefore there is a limit to engine efficiency (the ratio of how much work we get relative to how much energy we put in). All internal energy, or all heat cannot be converted into mechanical work. Only a part of these energy forms can be transformed into mechanical work. For instance, in coal-fired thermal power plants typical high efficiencies are about 45%. The remaining part of the supplied energy *must* be rejected by the environment.

In thermal processes, a *working substance* is compressed and/or expanded and receives and/or rejects heat from its surroundings. The changes in the state of the working substance may occur in a way so that one thermodynamic variable remains constant: Keeping the volume constant refers to *isochoric* processes; keeping the pressure or the temperature constant refers to *isobaric* and *isothermal* processes, respectively; finally, keeping the entropy or enthalpy constant refers to *isentropic* and *isenthalpic* processes, respectively. If heat transfer does not take place during a process, this is called an *adiabatic* process.

When heat is added to a substance, the internal energy of the working substance increases, which is what happens when we heat liquid water causing it to vaporize. This increase of internal energy can subsequently be used to generate work, which is what happens when the substance is expanded, e.g., in a turbine.

2.4 Energy and Electricity

Energy considerations are of great importance in electrical systems, interconnected circuit elements such as batteries, resistors, capacitors, and inductors (generally called two-terminal elements) as well as generators and motors, antennas, and various electronic equipment.

As an example of an energy system where the carrier of energy is "jingling" electrons in conductors, consider the circuit shown in Figure 2.3. The power absorbed by a circuit element having terminals A and B as shown in Figure 2.3 is simply

$$P = V \times I \tag{2.9}$$

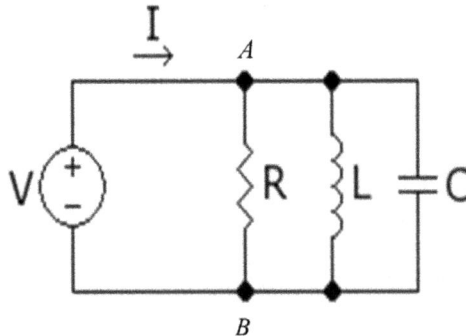

Figure 2.3. Idealized familiar components of an electric system.

If the power, voltage, and current quantities vary with time, that is, they are functions of time, the power relation above is written as

$$P(t) = V(t) \times I(t) \tag{2.10}$$

This is the instantaneous power absorbed by a circuit element or an entire circuit. To calculate the energy absorbed between times $t = t_1$ and $t = t_2$, we integrate the power over the time interval to obtain

$$E = \int_{t=t_1}^{t=t_2} P(t)dt \tag{2.11}$$

When a machine consumes power of 1,000 W = 1 kW over a time interval of 1 hour, we say that the machine has consumed a total energy of 1 kWh (1 kilowatt-hour). Conversely, when an electric power station produces 1000 MW (1000 megawatt) over a time interval of 1 hour, we say that the generator has produced 1 MWh (1 megawatt-hour).

It should be noted that economic transactions for electricity generation, consumption, or trading are done in energy units, such as kWh, MWh, or even terawatt-hours (TWh), not in power units.

A power station's turbine drives a generator producing the electricity delivered with the speed of light to any consumer, or load, connected to the power grid.

In electrical systems, efficiency is an important index for the ability of the system to transform heat into mechanical motion, which is the main mechanism for "jingling" electrons. An ideal heat engine receiving thermal energy from a heat reservoir at constant temperature T_{in} and rejecting heat at another reservoir having the temperature T_{out} operates with the maximum energetic efficiency given by

$$\eta = \frac{T_{\text{in}} - T_{\text{out}}}{T_{\text{in}}} \tag{2.12}$$

This equation was first presented by Sadi Carnot in 1824. The temperatures T are in the Kelvin scale. When the heat is supplied at varying temperatures, which is the case in practically all applications, thermodynamic average temperatures should be used in the above equation.

Modern steam power plants using coal (advanced ultra-supercritical coal, AUSC, at temperatures around 700°C) have efficiencies of around 45%, whereas combined-cycle power plants using natural gas can achieve efficiencies close to 60%, with to temperatures exceeding 1500°C). Seeking discoveries and innovations in advanced material that can withstand such high temperatures and go even higher is an active area of interest in generator technology as their efficiency is directly affected by their maximum internal temperature as given by Eq. (2.12).

In comparison to the above, a standard 1000 MW Pressurized Water Reactors (PWR) have efficiencies near 35%. Although PWRs are highly reliable baseload electrical generators of high-capacity factors, typically operating at 1000 MW, nonstop, 95% of the time, the fact that they operate at lower temperatures, (roughly 300°C) reduces their overall thermodynamic efficiency in accordance with Eq. (2.12). The temperature constraints in these reactors are based on thermal-hydraulics limitations which were empirically established several decades ago reflecting an early state of knowledge between fission, steam, steel and heat utilization to produce kinetic energy in the electricity generators.

2.5   Hydrocarbons and Energy Transition

There are different fuels or, sources of energy, for different uses that include, but are not limited to, the generation of electric power, transportation, industrial applications, and heating and cooling for residential, commercial, and industrial users.

Hydrocarbons, petroleum, coal, and natural gas are the principal sources of fuel. Since they are derived from plants that flourished millions of years ago, they are called "fossil fuels." Concerns about global climate change and hydrocarbon availability are motivating the search for alternative energy sources, a process often referred to as *green energy transition*, due to its emphasis on renewables (wind, solar, biomass, geothermal, and hydro). We have seen, in Chapter 1, that this is simply part of the old and ongoing Promethean energy transition.

Petroleum, or simply "oil," has been the world's leading hydrocarbon. Petroleum accounts for approximately 31% of primary global energy use. It is the fuel that enabled modern industrial developments and since the early 1900s has figured prominently in all geopolitical and economic conflicts, strategies, war, and peace.

Other hydrocarbons used for the production of energy are coal and natural gas. Coal is responsible for approximately 27% of global energy use and natural gas for about 24%. Coal is principally used for the generation of electric power but a growing share of power generation is shifting to natural gas due to concerns about climate change and growing gas availability, especially in North America, where new extraction technologies such as hydraulic fracturing or "fracking" are successfully applied to use shale, sands and coal bed methane fields. In addition to its growing role in power generation, natural gas remains an important feedstock for many chemicals, in a variety of industrial and agricultural applications (fertilizers).

All hydrocarbons, petroleum, coal, and gas, account for more than 82% of the global primary energy supply. The remaining primary energy is supplied by nuclear, biomass, hydroelectric, solar, wind, geothermal, and solar energy.

The mix of global energy as illustrated in the schematic diagram of Figure 2.4 shows how primary energy sources are apportioned in different domains of use. The percentages reflect average values over a decade. It should be noted that a major fraction of petroleum is used to cover transportation needs. It should also be noted that less energy is being used to produce electricity than to satisfy all other needs such as transportation, heating, and industrial uses. Since more than 82% of the total energy supply occurs by burning hydrocarbons, carbon emissions, and climate-change risks are associated with broader energy utilization than power generation. Hence, the focus on power generation as a carbon source invites non-trivial discussions on future energy mixes, particularly for the gargantuan energy needs of the AI infrastructure and the electrification of transportation in the future. We recall that renewables and nuclear power are almost exclusively used to produce electricity with a smaller carbon footprint (they both need hydrocarbons for construction and fabrication purposes, but less so during operations).

Example 2.2. Assume the average daily caloric intake per person is 2500 kcal. The global population is approximately 8 billion people.

(a) Calculate the total energy consumed by all people on the planet per day in Joules.

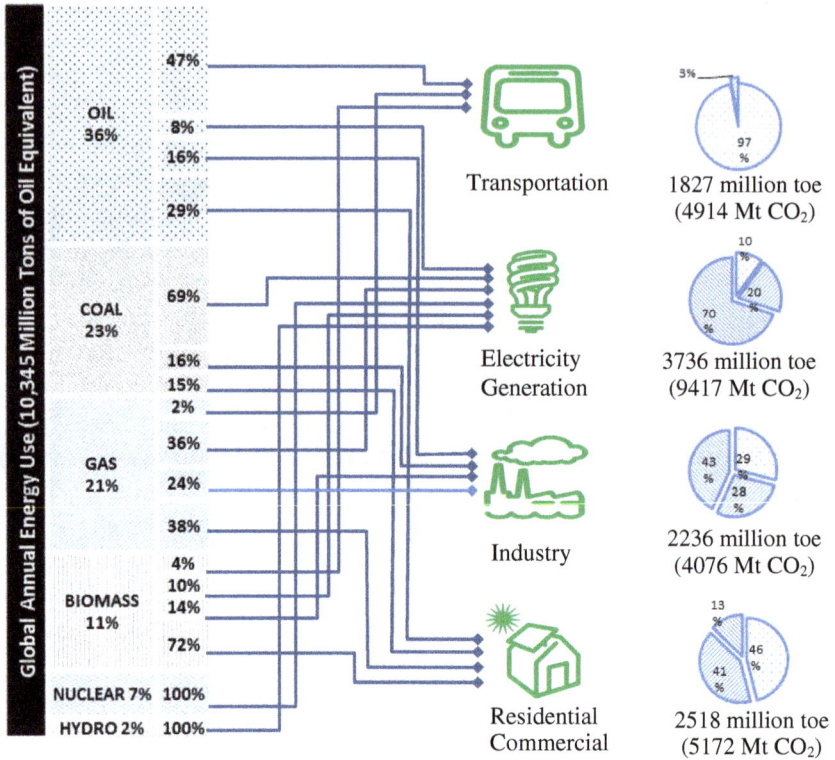

Figure 2.4.　The general breakdown of global annual primary energy use.

(b) If the total energy from food was provided by burning natural gas (with an energy density of 50 MJ/kg), calculate how much natural gas would be required per day.

Solution: The problem is indicative of energy modeling used to address questions in energy economics.

(a) Given that 1 kcal = 4.184 J, the energy consumed by 1 person per day is

$$2{,}500 \text{ kcal} \times 4.184 \text{ J/kcal} = 10{,}460 \text{ J} = 10.46 \text{ kJ}$$

If we have 8 billion people on Earth, the energy requirement to feed all of humanity in a day is

$$(10.46 \text{ kJ/person}) \times (8 \times 10^9 \text{persons}) = 83.68 \times 10^9 \text{ kJ}$$

This is equivalent to 0.08368 EJ (exajoules) per day, since

$$1 \text{ EJ} \equiv 10^{18} \text{ J}$$

Hence, the energy needed to fully feed all of humanity in a day is

Total food energy per day for all humans = 0.08368 EJ

And in an annual basis,

$$\text{Annual food energy} \approx \left(0.08368 \ \frac{\text{EJ}}{\text{day}}\right) \times 365 \text{ days} = 13.4 \text{ EJ}$$

It should be noted that 13.4 EJ is the total energy value of food consumed.

If all of this food energy was produced by modern agriculture, it is estimated that the actual energy used to produce and distribute this energy as food is approximately 12 times the caloric value, hence the total energy that would be needed to feed 8 billion people is approximately,

$$\text{Total energy to feed humanity} \approx 161 \text{ EJ}$$

This is a significant fraction of the total energy used annually by humanity which is estimated to be 620 EJ (EIA, 2025).

(b) The energy density of natural gas is 50 MJ/kg. The mass of natural gas is therefore equal to

$$\frac{83.68 \times 10^9 \text{ MJ}}{50 \text{ MJ/kg}} = 1.67 \text{ Mt (megatonnes)}$$

2.6 Hydrocarbon Supply Security

Hydrocarbons have fueled the unprecedent human progress over the last century and a half but have also been at the epicenter of war, strife and geopolitical re-alignments (Yergin, 2008). Beginning with a period of rivalry between the British and Russian Empires in the 19th and early 20th Century, control of energy resources figured prominently in both World Wars as well as in several local and regional conflicts since then. Dr. A. M. Bakhtiari (1946–2007), a prominent oil engineer, painted a picture of oil resources in the Middle East, which

Figure 2.5. Nearly 70% of global easily accessible petroleum reserves are in five countries of the Persian Gulf (within the horseshoe area).

is illustrated in Figure 2.5. Nearly 70% of the world's conventional oil reserves are located in five countries of the Persian Gulf in the area covered by the horseshoe shape. This is the underlying geopolitical importance of the Middle East and North Africa (MENA) for the modern world and lies at the heart of tensions, conflicts and wars.

The energy industry is one of the world's enabling industries. As an infrastructure, energy is the *infrastructure of infrastructures.* It enables the survival of humanity, the workings of industry and agriculture, the generation of progress, shelter, education, health and prosperity. Many crucial infrastructures including water systems, AI, cultivation and distribution of food, healthcare, transportation, maintenance, and dissemination of modern living standards are depending on the energy industry.

As a *thought experiment,* let us suppose that all supplied energy is derived from petroleum. A relative sense of production and consumption magnitudes is obtained through the concepts of a "barrel of oil equivalent (boe)" and "ton of oil equivalent (toe)," where,

$$1 \text{ metric ton of oil equivalent (toe)}$$
$$= 7.5 \text{ barrels of oil equivalent (boe)}$$

If the world's thirst for energy were to be satisfied solely on petroleum it would require approximately 270 million barrels of oil equivalent (boe) daily (2024 estimate).

Energy, however, is not fairly available to all humanity. For nearly two billion people (out of the eight billion world population) there is no significant access to modern energy forms. On the other hand, energy use by the world's rich countries varies considerably, with the US leading the world in per capita energy consumption (65 boe energy per person per year) followed by Japan and Germany (approximately 32 boe energy per person per year) and the UK (30 boe energy per person per year).

During the last 100 years, there has been persistent growth in global energy demand. On a *per-country* basis, satisfying energy demand has been the foundation of building modern economies and lifestyles. Energy availability is a prerequisite for building hospitals, schools, roads, airports, and numerous complex civil infrastructures. Energy has a significant contribution to rising standards of living and life expectancies, rising comfort, and unprecedented personal mobility through automobiles and airplanes. Yet, there are great margins for growth in energy appetites given that about 12% of the world uses 54% of the world's energy and a little less than a third of the world still has little access to modern energy.

World petroleum demand is at close to 103.1 million barrels of oil per day (mb/d) in 2024 and its growth potential is trending up (DOE/EIA, 2024). On the other hand, global oil production is very close to that in daily mb/d and although flat and down-trending in many countries of the world, it is moving up in the United States, largely due to the shale oil revolution. It should be noted that this includes crude oil production as well as other liquid fuels, suggestively called "substitute liquids," such as those derived from natural gas and oil sands.

The US Department of Energy's statistical agency, the Energy Information Administration (EIA), reported in 2024 that actual crude oil production in recent years remains rather flat around

102.6 mb/d despite high prices that, in theory, should be an incentive for production growth (DOE/EIA, 2024). Excess capacity in the global supply of light sweet crude oil appears to be very small, a troublesome sign pointing to fears of a phase transition in the world's oil markets as identified by King and Murray in a commentary in Nature some years ago (King and Murray, 2012).

Until the 1970s most of the world's oil transfers were achieved through bilateral contracts and government agreements. Since the early 80's competitive oil markets have become the place where the vast majority of the world's oil is traded. In addition to allocating oil, market mechanisms use complex financial instruments (e.g., derivatives) to allocate capital inter-temporally. Investments in new technology, infrastructure modernization, and future exploration plans, emanate also from risk-based decision-making trying to reduce volatility in the markets.

In recent years, diminishing excess capacities have posed a challenge to oil markets. At least 5% excess capacity is needed to have stable functioning markets. The conundrum posed by geological constraints in oil production in the face of reduced excess capacities threatens price elasticity and the market system as a whole (King and Murray, 2012).

A country's per capita oil consumption provides a telling story of its economic and political life as shown in Figure 2.6. In the early 1900s, the US embarked on building a modern economy that raised the per capita oil consumption from approximately 1 barrel per person in 1900 to a peak consumption of nearly 30 barrels per person in the early 1970s and eventually plateaued at about 25 barrels per person (after the two oil crises of the '70s). Similar patterns appear for Japan and South Korea, which came to a lower plateau of approximately 15 barrels per person per year by the late 1990s.

Astonishing growth in global oil demand however came recently from the rapid transition of China and India to more energy-demanding consumer economies. China and India currently consume nearly 3 and 1 barrels of oil per capita per year respectively. Their present course of development might reach a consumption level of 10–17 barrels per capita per year in 10–20 years. This is an important consideration for the world's oil markets because then the equivalent of several Saudi Arabia's oil reserves would need to be added to the global oil supply to meet this level of new demand.

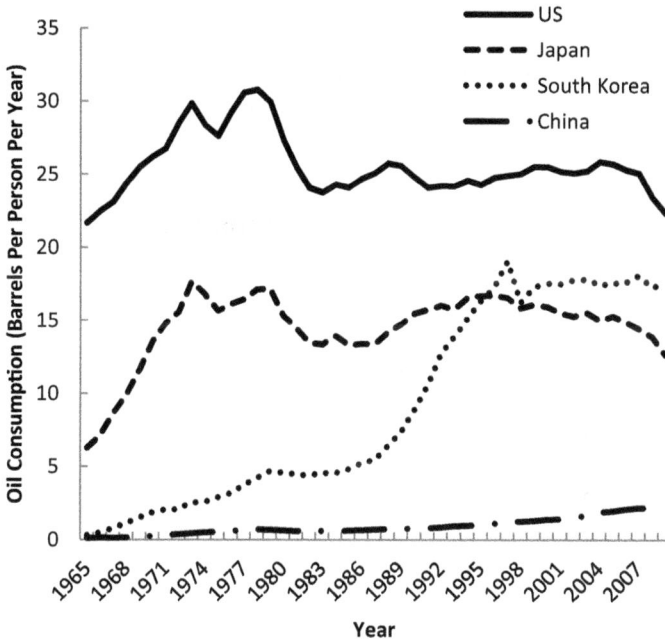

Figure 2.6. Oil consumption per capita since 1965.

In a way, part of the supply and demand imbalances in the oil markets was caused by the lack of anticipation and by the confusion that followed the Soviet Union's collapse in the early 1990s. At that period, reduced Russian demand for oil made nearly 8 mb/d available for exports in a rather short time. This created a sense of complacency regarding resource availability due to rapidly growing Russian energy exports.

Example 2.3. Suppose there are 1.4 billion cars in the world. On average, each car drives 15,000 km/y, consuming 8 l of gasoline per 100 km. The energy density of gasoline is 34.2 MJ/l. Assume that electric cars consume 0.2 kWh/km.

(a) Calculate the total annual energy consumption for all cars if they are gasoline-powered.
(b) Calculate the total annual energy consumption if 50% of the cars are replaced with electric cars.

Solution:

(a) The fuel consumed by each car per year is equal to

$$15{,}000 \text{ km} \times \frac{8 \text{ l}}{100 \text{ km}} = 1{,}200 \text{ l}$$

This translates to an annual energy consumption of

$$1{,}200 \text{ l} \times 34.2\frac{\text{MJ}}{\text{l}} = 41{,}040 \text{ MJ}$$

Therefore, the total energy for all cars would be equal to

$$1.4 \times 10^9 \times 41{,}040 \text{ MJ} = 57.456 \text{ EJ (exajoules)}$$

(b) The annual energy consumption of an electric car can be calculated as

$$15{,}000 \text{ km} \times 0.2\frac{\text{kWh}}{\text{km}} = 3{,}000 \text{ kWh}$$

Since 1 kWh = 3.6 MJ, the annual energy consumption of an electric car would be equal to

$$3{,}000 \times 3.6 \text{ MJ} = 10{,}800 \text{ MJ}$$

Therefore, the total energy for 0.7 billion electric cars would be

$$0.7 \times 10^9 \times 10{,}800 \text{ MJ} = 7.56 \text{ EJ (exajoules)}$$

And finally, the total energy for a mixed fleet of gasoline-powered and electric cars will be

$$28.728 \text{ EJ} + 7.56 \text{ EJ} = 36.288 \text{ EJ}$$

Compared to (a), there is a reduction in energy consumption of approximately 36.8%.

Example 2.4. A country consumes 1 billion barrels of oil equivalent (boe) annually.

(a) Convert this energy into tons of oil equivalent, given that 1 boe = 0.136 toe.

(b) Calculate the total energy consumption in Joules, knowing that 1 toe = 41.868 GJ.

Solution:

(a) The total energy in toe is

$$10^9 \text{ boe} \times 0.136\frac{\text{toe}}{\text{boe}} = 136 \text{ million toe}$$

(b) The total energy in Joules is equal to

$$136 \times 10^6 \text{ toe} \times 41.868\frac{\text{GJ}}{\text{toe}} = 5.694 \text{ EJ (exajoules)}.$$

2.7 Hydrocarbons and Climate Security

Climate security is an area of high media and political visibility and is strongly tied to fears of the Promethean energy transition reaching a dangerous juncture threatening the Earth's ecosystem. When all is said and done it may very well be that climate security is energy security in disguise; both share converging long-term goals, that is, limiting the use of hydrocarbons. Not surprisingly, leadership for mitigating possible climate change impacts has been provided by hydrocarbon-poor Europe and Japan while hydrocarbon-rich countries react less enthusiastically to the idea of decarbonization.

Climate Science indeed shows interesting variations in climate statistics indicating evidence for some climatic instability and unpredictability. There is very little apocalyptic in this evidence and the variations. The Earth's climate has always change and the record shows great resilience in absorbing variabilities and coping with large effects.

Nonetheless, potential risks to ecological balance and the economic health of humanity need to be carefully considered. They include food production, biodiversity, increased droughts and wildfires as well as migration and perturbed population dynamics. There is some controversy about the interpretation of data, but the scientific consensus points to anthropogenic causality in the phenomena, at least in part. Yet, there is little agreement on what mitigation strategies to adopt to address the risk that the phenomena may pose in the future (Lomborg, 2020).

Consider, for example, reported evidence that 95% of land-based glaciers are losing mass. This is an area that has been controversial due to issues associated with data for the Himalayan glaciers. In an earlier study focusing on alpine glacier statistics, a 50-year window of negligible volcanic emissions in the Northern Hemisphere was identified during which period alpine fluctuations should be balanced on average (Ott and Tsoukalas, 2001). Surprisingly, alpine glaciers continued to retreat during this time and the principal causal agency was increasing CO_2 in the atmosphere.

Consider Figure 2.7 showing a window of time (1928–1964) with negligible changes in the earth's optical depth (in the middle of the figure). This signifies the absence of particulates and sulfuric emissions that are released from major volcanic eruptions and counter global warming, in effect they provide a shield from incoming solar radiation. The nearly three-decade-long window offers "laboratory conditions" in Nature to study the fluctuations of 300 Alpine glaciers for which Swiss authorities and glaciologists have very good statistics collected during the past couple hundred years. Atmospheric sulfate aerosols of volcanic origin materially affect glacier fluctuation. Strong glacier advance was initiated during the high-aerosol periods 1902 to 1914 and 1963 to 1985.

The advance/retreat statistics is shown in Figure 2.8 where the dominance of retreat over advance is seen. Between 1915 and 1964

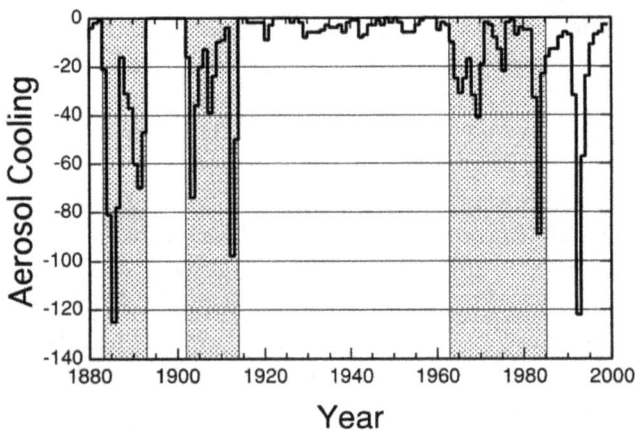

Figure 2.7. Optical depth of the atmosphere.

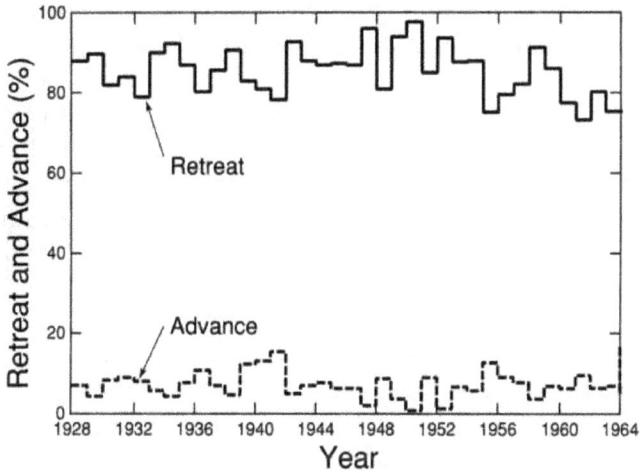

Figure 2.8. Advance vs. retreat of Alpine glaciers during a period of low volcanic emissions.

there was an almost 50-year "window" of negligible aerosols during which the underlying radioactive forcing can be evaluated. The underlying forcing reversed the ongoing advance of the majority of the 300 larger Alpine glaciers and established a dominance of retreating over advancing glaciers by more than a factor of ten: 86% (±6%) of the glaciers were in retreat and only 7% (±3%) advanced.

The analysis of causal factors by Ott (Ott and Tsoukalas, 2001) points *ceteris paribus* to growth in CO_2 concentrations as the most likely cause for the Alpine glacier retreat during this period. Generally, there is likely an anthropogenic component to the observed global temperature rises, although there is less agreement on what consequences may be there, and what to do about it. But it is hard to tell if this is a local or indeed global effect.

For instance, during the period of classical Rome (circa 100 BCE), temperatures in the Mediterranean are estimated to be 2°C higher than current temperatures. This is known as the Roman Warm Period (250 BCE to 400 CE), but like other pre-industrial warm or cold periods, it might have been a regional phenomenon. In contrast, current global warming appears to be more severe and widespread, with global average temperatures likely to exceed 1.55°C above preindustrial levels in 2024.

The Paris Agreement is the result of the United Nations Climate Change Conference (COP21) which entered into force in 2016. The Agreement aims to limit global temperature rise and reduce overall greenhouse gas emissions to keep the increase in global average temperature to well below 2°C above pre-industrial levels, and preferably limit the increase to 1.5°C. Emissions of greenhouse gases, including CO_2 are expected to be reduced. Yet, the Paris Agreement's goals may be hard to meet. The world is currently on track for a temperature rise of 2.5–2.9°C above pre-industrial levels this century. To align with the 1.5°C goal, global greenhouse gas emissions need to be cut by 42% by 2030 and 57% by 2035.

These goals appear antithetical to reducing global energy poverty. In addition, likely mitigation strategies to limit temperature rises appear too unrealistic to be implemented. Instead, a measured approach may involve emissions limitations together with the growing availability of emissions-free, dispatchable sources of affordable electricity such as nuclear power including small modular reactors (SMR) and microreactors (MR). Embracing the Atomic energy transition on a global scale may thus provide emissions-free and affordable decarbonized electricity (not energy in general) which will be needed for Humanity to mitigate the potential consequences of climate change in realistic terms.

The efforts by industrial economies to meet these temperature limits through a combination of reduction measures and international or regional mechanisms have led to the view that a metric ton of CO_2 (a carbon unit) is a tradable commodity. An equivalent ton of CO_2 can be used to measure the effects of emissions of other greenhouse gases.

Carbon units exchanged in carbon markets are bound to add considerable and varying costs to energy utilization. Whether they take the form of allowances (paying for the "right to pollute") or carbon credits (also known as carbon offsets or emission credits) they affect the cost of supplying energy and thus may incentivize technologies and approaches for using energy derived from hydrocarbons more efficiently.

It is worth mentioning that in the so-called cap-and-trade schemes that are based on an initial allocation and subsequent trade of carbon allowances, tougher emission caps and carbon prices are positively linked. A new model has emerged for managing fossil fuel utilization. This model complements the traditional supply and demand model.

It prioritizes decarbonizing electricity through the Atomic energy transition while using natural gas to get there sometime before the end of the 21st Century.

Problems

Problem 2.1. A 2-kg pendulum bob is attached to a string 2 m long. The pendulum is initially released from rest at an angle of 30° from the vertical. Assume no air resistance.

(a) Calculate the kinetic energy of the pendulum bob when it passes through its lowest point.
(b) Calculate the potential energy of the pendulum bob relative to its lowest point when it is at the highest point in its swing.
(c) Verify that the total mechanical energy remains constant throughout the pendulum's motion.

Problem 2.2. A circuit consists of a 12-V battery connected to a resistor with a resistance of $R = 4 \, \Omega$.

(a) Calculate the power dissipated in the resistor.
(b) If the circuit operates for 5 min, calculate the total energy dissipated in the resistor.

Problem 2.3. A city requires 1×10^{15} J of energy per year to meet its electricity demands. This energy can be supplied by different sources: (i) coal, with energy density $= 24$ MJ/kg; (ii) natural gas, with energy density $= 50$ MJ/kg; and (iii) solar panels, where the average power output per panel is 300 W, with six effective sunlight hours per day.

(a) Calculate the mass of coal required to meet the city's energy needs for a year.
(b) Calculate the mass of natural gas required for the same energy demand.
(c) Determine the number of solar panels needed to meet the energy demand.

Problem 2.4. A city plans to switch all its vehicles to electric vehicles (EVs), requiring a total energy of 5×10^{12} J/y to charge them. The energy can be provided using different sources: (i) wind turbines, where each turbine produces 2.5 MW on average and operates for 40% of the year; (ii) hydropower, where a dam generates 50 MW continuously; and lastly (iii) batteries powered by diesel generators, which provide energy at 45 MJ/l.

(a) Calculate how many wind turbines would be required to meet the energy demand.
(b) Calculate how many hours the hydropower plant must operate to meet the energy demand.
(c) Determine the total volume of diesel fuel required if the energy is provided entirely by diesel generators.

Problem 2.5. The energy demand of a small country is 2.5×10^{15} J/y. One metric ton of oil equivalent (toe) represents about 42 GJ of energy.

(a) Calculate the total energy demand of the country in terms of toe.
(b) If the country's energy demand is met using natural gas (with an energy density of 50 MJ/kg, determine the mass of natural gas required in metric tons.
(c) Compare the energy from 1 toe to the energy produced by a 1 MW wind turbine operating for 2000 hours in a year. How many wind turbines would be required to produce the same energy as 1 toe?

References

Halliday, D. and Resnick, R., *Physics Parts I and II*, John Wiley & Sons, New York, NY, 1966.

King, D. and Murray, J., "Oil's tipping point has passed," *Nature*, 481, 433–435, 2012.

Lomborg, B., *False Alarm: How Climate Change Panic Costs Us Trillions, Hurts the Poor and Fails to Fix the Planet*, Basic Books, New York, NY, 2020.

Ott, K.O. and Tsoukalas, L.H., "Alpine glacier fluctuations, volcanic aerosols and radiative forcing," *World Resource Review*, 13(4), 461, 2001.

US Energy Information Administration (EIA), *Annual Energy Outlook 2025*, https://www.eia.gov/outlooks/aeo/ April 15, 2025.

Yergin, D., *The Prize: The Epic Quest for Oil, Money and Power*, Free Press, NY, 2008.

Chapter 3

The Atomic Energy Transition

3.1 Extracting Energy From Atoms

AI and the Atomic energy transition, symbiotically, promise to be the most consequential undertaking of Humanity going forward into the future.

The 20th-century extraordinary discoveries of Modern Physics paved the way for the Atomic Energy Transition. During the late 1930s, German scientists Otto Hahn, Lise Meitner, Fritz Strassmann, and Otto Robert Frisch, discovered that uranium nuclei could be disassembled, or fissioned, by neutrons. It soon became apparent that there was significant energy release, and that more neutrons were released (typically two to three for each neutron absorbed and causing fission). Hence, a chain reaction is possible, with neutrons released from one fission triggering more fissions, leading to a sustainable release of energy. Soon thereafter it was established that neutrons produced can also be used to create new fuel (in a process called breeding).

Figure 3.1 illustrates the fission process. Energy is released according to Einstein's famous formula $E = mc^2$, where m is the matter transformed into energy and c is the speed of light ($c = 300{,}000$ km/sec). For instance, in the case of fissioning uranium, the energy released is equivalent to the mass of a neutron which is approximately 200 MeV (measured in MeV units of energy), that is, 200,000,000 eV. By contrast, combustion releases eV scale energy for

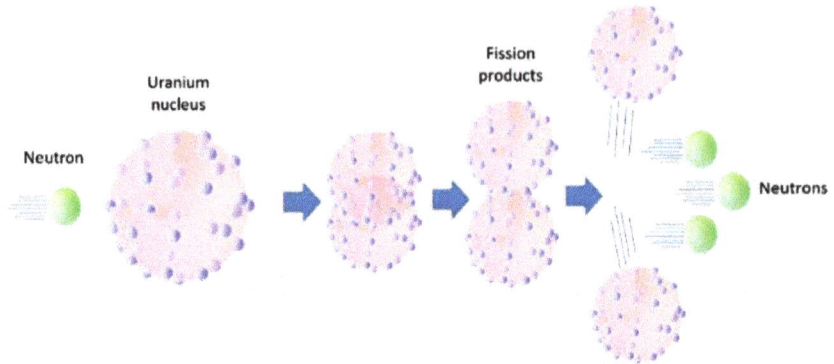

Figure 3.1. The disassembly, or fissioning, of $^{235}_{92}$U (left), produces fragments with high kinetic energy, i.e., heat (right), but also more neutrons.

each methane molecule disassembled (as shown in Chapter 2). This is an important point to note: Disassembling atoms results in millions of times more energy than disassembling molecules.

Machines that extract energy from atoms are called "reactors." There are two types: *fission reactors* if the method of harvesting is disassembling atoms, *fusion reactors* if the energy is extracted from joining atoms (fusing them) to produce new atoms and obtain energy.

Unlike Promethean energy, with atomic energy sustainability is built in. The chain reaction can create its own fuel as it produces energy and the more energy it produces, the more fuel it makes available for future energy production. This is a process known as "breeding" and the machines to do it are called "breeder reactors." An analogous process holds true for nuclear fusion, where scarce tritium can be bred from abundant deuterium. In this chapter, we will examine what reactors are and how they can produce energy and fuel at the same time.

The development of reactor technology in the decades after the end of World War II, preceded with development of a couple dozen different reactor types. These were the first decades of the Atomic Age (1942–1980). After a hiatus of nearly forty years, several new designs are emerging recently, focusing on smaller reactors including Small Modular Reactors (SMR) of less than 300 MWe power and Microreactors (MR) of less than 10 MWe.

In dealing with atoms, it is important to remember that the number of atoms in a typical quantity of matter (a mole) is very large.

Atomic quantities are comparable to Avogadro's Number that is, $N_A = 6.022 \times 10^{23}$ atoms per mole. Hence, our approaches are statistical and when we talk about an atom, a nucleus, or a neutron or proton, we are essentially talking about representatives of large numbers of particles — in the order of Avogadro's Number.

A modeling assumption in our formulations is that a nucleus that could be disassembled, say $^{235}_{92}\mathrm{U}$, is stationary while an incoming neutron is moving as shown in Figure 3.1.[1] The probability that the two, stationary nucleus and moving neutron, will interact with each other is a function of the kinetic energy of the incoming neutron and is measured in units of a very small number, called "barn," defined as

$$1 \text{ bn} \equiv 10^{-28} \text{ m}^2 = 10^{-24} \text{ cm}^2$$

The notion of probable interaction is called "microscopic cross-section," it is given in units of area analogous to a target in a "bull's eye diagram," and is indicated by lowercase sigma, that is

$$\sigma \equiv \text{nuclear area seen by a neutron for an interaction}$$

One such type of interaction is fission as shown in Figure 3.1. But there are other interactions, in addition to fission. They are categorized in two broad categories: *absorption* or *capture*, and *scattering*. All have their own microscopic cross sections, and thus a hierarchy of cross sections can be formed reflecting the sequence of possible subatomic interactions and which can be linearly added to each other. For instance, since all interactions are categorized as either "scattering" or "absorption," the sum of these two constitutes the *total* microscopic cross section, i.e.,

$$\sigma_T = \sigma_s + \sigma_a \tag{3.1}$$

---

[1]The standard notation for isotopes is $^A_Z X$, where X is the chemical label for the element, say U for uranium, A is the "Mass Number," that is the sum of protons and neutrons in the atomic nucleus, and Z is the "Atomic Number," that is the number of protons. The Atomic Number Z also gives the chemical identity of the atom, and hence can be omitted. For instance, when $Z = 92$ the element is Uranium and for this reason if the element is listed explicitly, 92 is omitted. The Mass Number A gives the identity of the isotope and in the nuclear discourse it is important to list it. For instance, $^{235}_{92}\mathrm{U}$ refers to the isotope of uranium that has 92 protons and 235 protons plus neutrons in its nucleus. It is common to omit the Atomic Number 92 and list the isotope simply as $^{235}\mathrm{U}$.

where

σ_T is the total microscopic cross section,
σ_s is the scattering microscopic cross section, and,
σ_a is the absorption microscopic cross section

But the microscopic scattering cross section can be considered as the sum of all scattering interactions which also fall into two categories: *elastic* or *inelastic*. Hence,

$$\sigma_s = \sigma_e + \sigma_{\text{in}} \qquad (3.2)$$

where

σ_s is the total scattering microscopic cross section,
σ_e is the elastic scattering microscopic cross section, and,
σ_{in} is the inelastic scattering microscopic cross section

Similarly, all absorption interactions can be either interactions related to the disassembly or fission of the interacting nucleus, or simply capture without fission. The interaction that leads to fission is treated as part of all absorption interaction, since a neutron must be first absorbed by the nucleus — causing a major "stir" or instability inside the nucleus resulting almost instantaneously in the disassembly or fission of this nucleus. However, not all absorptions lead to fission. Other interactions may take place inside the nucleus. Hence,

$$\sigma_a = \sigma_f + \sigma_c \qquad (3.3)$$

where

σ_a is the total absorption microscopic cross section,
σ_f is the fission microscopic cross section, and,
σ_a is the capture microscopic cross section.

The capture microscopic cross section, in turn, may result is several interactions of a neutron with a nucleus that may produce a photon, or gamma ray (γ), or may an alpha ray (α), or even a proton (p). Hence,

$$\sigma_c = \sigma_\gamma + \sigma_{(n,a)} + \sigma_{(n,p)} \qquad (3.4)$$

where

σ_c is the total capture microscopic cross section,
σ_γ is the capture with γ emission microscopic cross section,
$\sigma_{(n,a)}$ is the capture with α emission microscopic cross section,
$\sigma_{(n,p)}$ is the capture with p emission microscopic cross section.

A fission interaction requires that a neutron is first captured by a nucleus before the nucleus breaks apart. Hence, fission follows absorption. Again, not all absorptions lead to fission. Beta decay may transform a neutron inside a nucleus into a proton thus changing the atomic number, Z, which is the total number of protons in the nucleus and hence the total number of electrons surrounding the atomic nucleus. The atomic number gives the atom its chemical identity.

On the other hand, if a neutron scatters before it gets absorbed, there is no chance for fission. This is a kind of free-lancing neutron that may be available for breeding new fuel, or it may be lost by leaking out of the core volume of a reactor. Again, the key idea is that all these interactions are based on probabilities or likelihoods involving large numbers of particles.

Figure 3.2 illustrates the varying likelihood for fission in a situation where a neutron of certain kinetic energy is approaching a ^{235}U or a ^{238}U nucleus. The neutron is characterized by its kinetic energy (measured in MeV and displayed in a logarithmic scale on the *x-axis* of the figure). The likelihood of interaction is displayed in barns, again on a logarithmic scale, at the *y-axis*. The fission interactions of neutrons of different energies with nuclei ^{235}U or ^{238}U are shown in the blue curves labeled "235 fission" and "238 fission" respectively. The interactions that lead to absorption without fission are shown in the red curve labeled as "238 capture." We note that the probability of fission increases significantly for ^{235}U when the incoming neutron has low energy, that is, it is approaching slowly. The probability of fission for low neutron energies approaching ^{238}U is zero. Only when a neutron has very high energy, roughly in the 1 to 10 MeV range, is the fission of ^{238}U somewhat probable as indicated by the curve in the lower right corner of Figure 3.2.

The isotope of ^{238}U constitutes 99.3% of natural uranium while the isotope ^{235}U is only about 0.7%. Yet, despite the low abundance

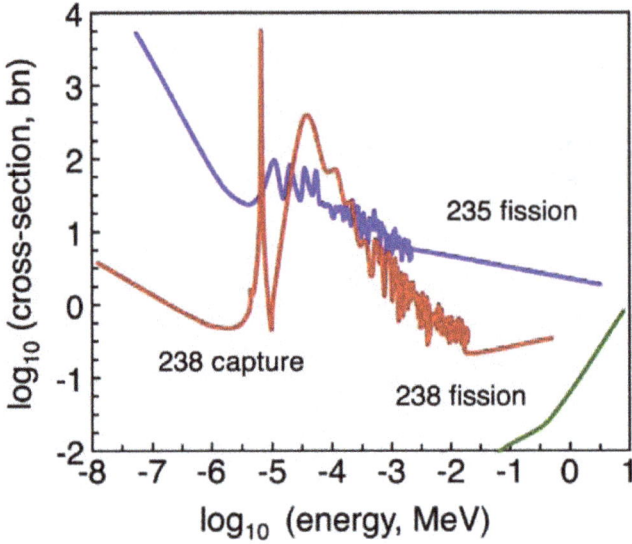

Figure 3.2.   Fission cross-sections of ^{235}U and ^{238}U as well as neutron capture cross-section for ^{235}U, all as functions of neutron energy.

of ^{235}U, in natural uranium the probability of fission outweighs neutron capture for neutrons with energies below approximately 0.1 eV.

Natural uranium is mostly ^{238}U. The likelihood that a neutron is captured (absorbed) by ^{238}U is significant across different energy (the red curve in Figure 3.2) but it does not lead to fission. Instead, it metamorphoses ^{238}U to a new isotope called Plutonium-239, denoted as $^{239}_{94}$Pu, which is fuel of great quality very similar to $^{235}_{92}$U, only better. The metamorphosis is not instantaneous, it takes a few days (as seen in Section 3.2). But, given the decades or even century-long operation of a reactor, we consider it instantaneous, hence a unique feature of the Atomic energy transition: energy and fresh fuel are produced concurrently.

Neutrons produced in fission have high kinetic energies (around 2 MeV), making a sustained chain reaction in natural uranium very unlikely. To overcome this impediment, neutrons must be slowed down through scattering interactions with other materials. This process, called "moderation," reduces their kinetic energy fast until they reach thermal energies, that is, they are in thermal equilibrium with a surrounding medium. Neutrons in very low kinetic energy are called "thermal neutrons." Their most probable energy is 0.025 eV (corresponding to approximately 2200 m/s neutron speed) at room temperature.

To achieve a successful chain reaction, the high kinetic energy neutrons produced in fission, as shown in Figure 3.1, must be slowed down to low energies called "thermal energies." This is achieved through a material called "moderator," which is typically water or graphite. Water can have a dual use. In addition to being a moderator, it can also be a coolant, that is, it can remove the heat produced by the fissioning of ^{235}U. For this reason, most reactors used for generating electricity use water for cooling and moderation and are called Light Water Reactors (LWRs). The adjective "light" refers to normal water and not "heavy water" which has deuterium bonded to oxygen in the molecule of water.

The considerations above have led to the development of two major types of reactors. First, *thermal reactors*, a category of reactors utilizing moderators (like water in Light Water Reactors or graphite in gas-cooled reactors) to slow down neutrons to thermal energies. Thermal reactors enable the use of fuels which are natural uranium slightly enriched in ^{235}U (typically less than 5%). Second, *fast reactors*, which is a class of reactors where the chain reaction is achieved and maintained using "fast" neutrons (high energy), requiring significantly enriched uranium fuel with a higher proportion of ^{235}U.

If we examine the detailed fission cross-section of ^{235}U shown in Figure 3.3, we observe three different regions across different neutron energies: *thermal, epithermal,* and *fast*. Neutrons are born in the fast region (in the right of the figure). But they need to lose energy and cross over the epithermal region before they reach the thermal region where the probability of causing fission increases exponentially (linearly on a logarithmic scale). Hence, neutrons must become thermal before they become productive. The middle region, the epithermal, is the wildcard region. Many nuclei, particularly heavy elements like uranium and plutonium, exhibit sharp *resonances* in their cross-sections over specific energy ranges (typically between

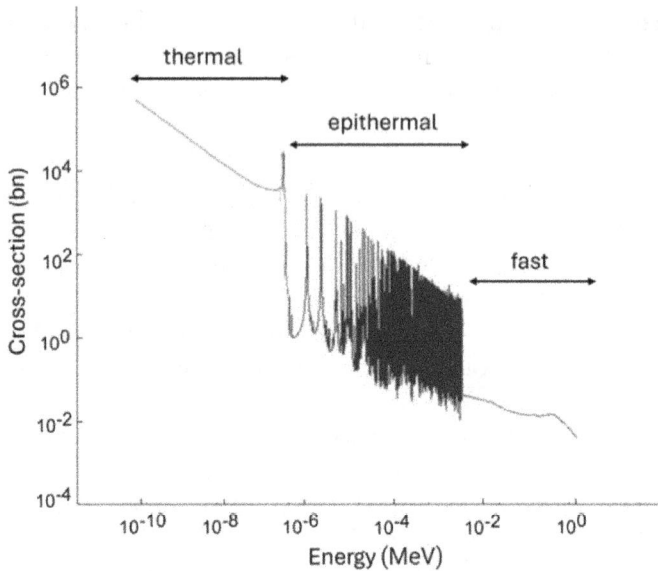

Figure 3.3. Fission cross-sections for the three regions of neutron energies, *thermal, epithermal* and *fast*.

1 eV and 100 eV). These resonances significantly influence neutron interactions and must be accurately accounted for in reactor calculations. Yet a detailed and deep quantum treatment is very difficult as the tools available are computationally very expensive and the theory likely inadequate to fully capture the complexities of Nature at the sub-neutron level (Economou, 2011).

For neutrons to trigger new fissions as they are born fast and must slow down to thermal, they need to get over this middle region. The reason is that in the epithermal region, neutrons can be absorbed without producing new fissions. The physics of the epithermal region are very complex, with absorption properties varying in ways not completely known but for engineering purposes sufficiently understood phenomenologically to be able to design fuel and moderator/coolant variables such as local (meaning space dependent) temperatures, flow rates, and neutron fluxes.

An issue of importance is the *Doppler effect*, which arises from the thermal motion of nuclei within the reactor fuel. This thermal motion causes the observed width of nuclear resonances to broaden, influencing neutron interactions. The capture rate within a resonance region is influenced by temperature. As temperature increases, the resonance peak becomes broader and slightly lower while the neutron flux at the peak decreases. This overall leads to an increase in the capture rate with increasing temperature. This temperature-dependent increase in capture rate contributes to a negative temperature coefficient of reactivity. This is because of increased capture in ^{238}U (a fertile material) which reduces the availability of neutrons for fission. While fission in ^{235}U or ^{239}Pu also increases with temperature, the increase in capture reactions generally dominates. As we will see later on, a negative Doppler coefficient of reactivity is crucial for reactor safety, as it provides an inherent mechanism for reducing reactivity and stabilizing reactor operation during temperature excursions.

Generally, thermal, and fast reactors serve different purposes. The first dominate thermal reactors used for energy production while the second are more important in fast reactors useful for producing new fuel (while still producing energy). Thermal reactors use neutrons of low kinetic energies, roughly 0.025 eV. Fast reactors use neutrons born in the fission process with 4 MeV initial kinetic energy.

Both thermal and fast reactors need a *coolant* to successfully remove the enormous heat generated by the fission process (approximately 200 MeV per fission). In addition, thermal reactors need a *moderator* such as water to slow down the fast neutrons born of fission, to make them slow. In Light Water Reactors (LWR), the coolant and the moderator are one and the same. Figure 3.4 shows the characteristic structure of a thermal LWR. A reactor is a machine that produces a lot of heat inside fuel rods (or plates in some designs). The heat is removed by a coolant and turned into steam through a steam generator (boiler) which in turn produces mechanical work, that is, it turns the shaft of a generator to produce electricity. More than 65% of all power reactors worldwide are of this kind. They are called Pressurized Water Reactors, or PWRs. There are several variants of this design, but essentially, they share similar features in the manner that they produce steam to generate electricity. A less popular variant is the Boiling Water Reactor, or BWR, where steam is produced inside the pressure vessel; here there is no steam generator. Steam from the reactor flows directly to the turbines that drive the generation of electricity. Less than 20% of the world's power reactors are BWRs.

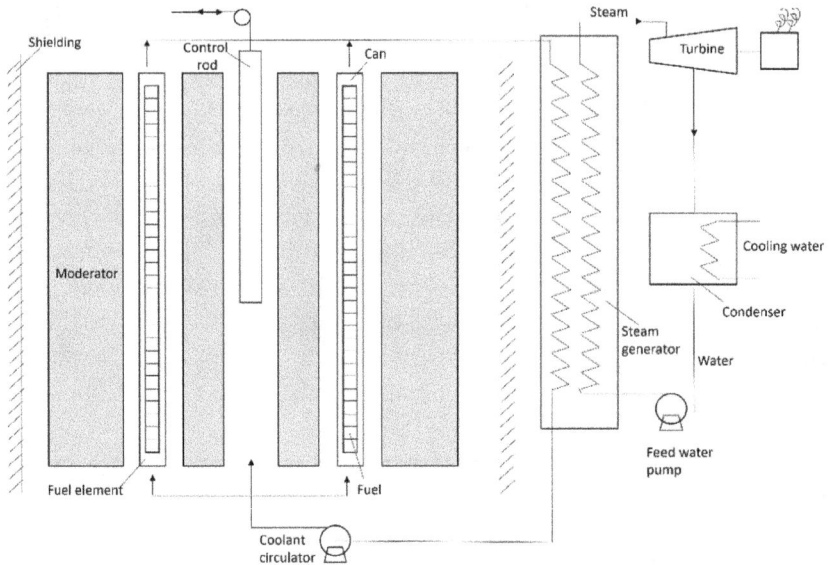

Figure 3.4. A thermal reactor needs a moderator to slow down neutrons before they can cause fission in a chain reaction and a coolant to remove the heat produced.

To achieve safe operations in producing electricity by disassembling atoms in a chain reaction, a reactor has several major components:

First, *the nuclear fuel.* The fuel is inside fuel rods aggregated in bundles called fuel assemblies. Uranium dioxide (UO_2) is a typical light water reactor (LWR) fuel, while uranium bicarbide (UC_2) is typical fuel for high temperature gas cooled reactors (HTGR). Each rod is approximately one cm in diameter and four meters long and holds some 200 to 300 fuel pellets of uranium oxide (typical PWR fuel). The pellets are stacked end-to-end within the fuel rod's cladding. The rods are bundled in lattices of 15×15, 16×16 or 17×17. The rods are hollow tubes made of special alloys (typically zircalloy) that can withstand large pressures, high temperatures, and high radiation. The behavior of the fuel is governed by interactions between neutrons, fissile material, fertile material and fission products which are modelled with a combination of quantum physics and phenomenological approaches that are partially satisfactory but remain incomplete.

This is an area where AI discoveries are expected to have a major role as we move forward with new reactor designs (Prantikos *et al.*, 2023, 2024).

Second, *the coolant.* Most thermal reactors use water as coolant, to remove the heat produced. A small number of reactors use gas such as helium and in the case of fast reactors coolants are typically liquid metals (such as an alloy of sodium and potassium NaK).

Third, *the moderator.* This is material of low atomic number whose purpose is to scatter neutrons inelastically so that they can lose a lot of energy through the smallest possible number of scatterings. Almost all light water reactors (LWR) use water as the coolant, but they also use the same water as moderator. In gas-cooled reactors (e.g., HTGR), the moderator is different from the coolant, typically graphite, while the coolant is gas. Carbon has a low atomic number; therefore, fast neutrons moderate, or loose, their energy to achieve energies that increase the likelihood of fission or absorption which leads to the breeding of new fissile material. Deuterium $_1^2D$, the isotope of hydrogen with one neutron and one proton is very stable and hence heavy water, D_2O, is an excellent moderator (used by the Canadian CANDU reactors). It should be noted that $_1^2D$, $_8^{16}O$, $_{16}^{12}C$ are all very stable isotopes because they have a nuclear shell structure that has the "magic numbers" for protons and neutrons. Magic numbers are 2, 6, 8, 14, 20, 28, 50, 82, 126. Especially when both atomic (Z) and mass number (A) are magic we have "doubly magic nuclear structures," which means that these nuclei have very low probability for neutron absorption, that is very small neutron absorption cross sections, a situation that is favorable to the overall neutron economy of a reactor (Lamarsh and Baratta, 2001). Since we need fast neutrons for fission, no moderator is needed in fast reactors.

Fourth, *special alloys.* These are structural materials holding fuel rods and fuel assemblies but also control rods used to control the chain reaction. Control rods have typically boron-based materials that absorb neutrons without leading to fission. They maintain the economy of neutrons for the chain reaction at a steady state. They can terminate the chain reaction if there is a threat to the reactor. The sudden termination of the chain reaction is called "scram."

And finally, concrete structures. The fifth major item is concrete structures that hold all other components within a building, called "containment, typically made of highly reinforced concrete protecting the reactor from outside threats, including missiles or airplanes. In addition, containment is the last barrier that protects the environment from any radioactivity that may be released in case of an accident. The concrete for nuclear structures has special specifications and is called "nuclear-grade."

It should be noted that reactors are machines; hence they may malfunction. The most severe category of malfunction is a *core meltdown*. A total or partial meltdown is not impossible. However, the design should be such that even if there is a core meltdown there should be no release of radioactivity to the environment. For this reason, the design should have multiple protective barriers and protection systems independent from control systems. The overall safety architecture provides *defense-in-depth* so that the probability of a severe accident leading to core damage should be no higher than 10^{-8} *per year*. This is a figure of merit which reflects a goal of no more than one meltdown per 100 million years of reactor operation.

Figure 3.5 shows a simplified schematic of a Pressurized Water Reactor (PWR) identifying the major components and systems. The chain reaction takes place in the *core* which is where the fuel is as well as the control rods and supportive structures. After a start-up maneuver of checks and gradual power ascension (it usually lasts several days) the reactor operates at its nominal power level, typically 1,000 MWe, in steady state, and non-stop for about 2 years (or up to 10 or more years for advanced designs).

The coolant (water) enters the pressure vessel at a temperature of approximately 290°C. After it goes through the core, it picks up heat produced by the nuclear chain reaction and exits the pressure vessel having temperatures around 330°C. The coolant then is transferred the steam generator (a boiler) where the heat is used to produce steam. Steam exiting the steam generator goes through a series of turbines turning electricity generators (not shown in Figure 3.5). After the steam has transferred most of its energy to the turbines, it condenses, that is it becomes liquid again, and is brought back to the pressure vessel.

In a Pressurized Water Reactor (PWR), the type that makes up more than 70% of the world's reactors, the water in the reactor is kept under constant pressure with the help of a Pressurizer, and thus the process of generating steam is done in a secondary loop (steam generator – turbines) with the primary loop (reactor – steam generator) having its own separate liquid (not mixing primary and

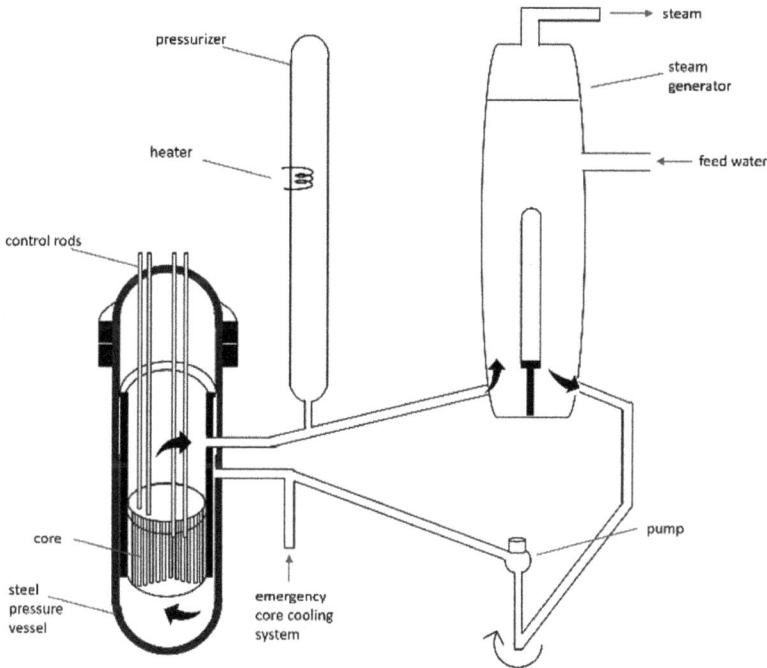

Figure 3.5. A typical PWR has one or more loops (only one shown here) with a steam generator (boiler) where the heat produced in the reactor core (within the pressure vessel) is used to make steam in the boiler.

secondary liquids limits the possibility of transferring any small quantities of radioactive material to the balance of plant, that is the generating station at large). Typically, older PWRs have two or three secondary loops, but most modern designs like the European Power Reactor (EPR) and the Russian VVER-1000 have four secondary loops. However, there are also recent designs like the American AP-1000 and the Chinese CP-1100 that have 2 secondary loops. For convenience, only one secondary loop is shown in the reactor schematic of Figure 3.5.

There is another type of LWR called Boiling Water Reactor (BWR), not as common globally, and in the almost canonical power level about 1,000 MWe. BWRs make up less than 15% of the world's reactors and their design has no secondary loop. Instead, steam is produced inside the pressure vessel and then directed to a series of turbines to generate electricity.

In total, some 400 power reactors, mostly PWRs and some BWRs have been deployed globally over the past 50 years, providing nearly 13,500 years of reactor operation with only three major accidents (Three Mile Island in 1979, Chernobyl in 1986, and Fukushima in 2011). It should be noted however that the worst accident was Chernobyl, which had an atypical dual-purpose reactor intended to produce power as well as weapons-grade plutonium. For this reason, Chernobyl type reactors (called with their Russian acronym RBMK) lacked inherent safety features such as "negative temperature coefficient of reactivity." In addition, in order to reduce construction costs RBMKs had no protective containment structure and during the accident major quantities of radioactive isotopes were released to the environment. Much radioactive material of the entire reactor core was actually sent to the troposphere and dispersed in large parts of Europe and Western Asia due to an extremely hot graphite fire burning at about 4,000 degrees centigrade and extraordinarily difficult to quench.

The isotopes that can act as nuclear fuel are called *fissile material*. There are three fissile materials: ^{235}U, ^{239}Pu, ^{233}U. The first is naturally occurring, albeit in small quantities (0.7% of natural uranium), but indispensable in starting a nuclear power program. The second is not found in nature, it is manufactured by nuclear reactors which need the first fissile material to get the process started. The third is also produced in reactors (from Thorium) as we shall see in the next section. While natural uranium contains both ^{235}U which undergoes fission with thermal and ^{238}U which primarily undergoes fission with fast neutrons, the low abundance of ^{235}U presents a significant challenge in achieving a self-sustained chain reaction. In a reactor, the abundant isotope of natural uranium, ^{238}U is more likely to capture or scatter neutrons, hindering the fission process but leading to the unique aspect of the Atomic energy transition, that is, the capacity to breed new fuel. New fuel in the form of the fissile material ^{239}Pu is produced from ^{238}U. Hence ^{238}U is called *fertile material*, meaning that from ^{238}U we can breed the fissile material of ^{239}Pu.

3.2   How Reactors Work

If we abstract away from the components of a typical reactor shown in Figure 3.5 and focus on the *core*, we can understand how reactors work. The core is the most important component, for that's where the disassembling of atoms takes place. Understanding what happens in the core is crucial for building an appreciation of the unique aspects of the Atomic energy transition.

The chain reaction is all about neutrons and in the reactor core the clear protagonists are neutrons. Neutron populations are agile and mobile whereas fissile isotopes are bulky and stationary (within thermal noise). Fissile isotopes interact with incoming neutrons including interactions that reproduce and rejuvenate the neutron population. Neutron energies and directions of travel are important attributes of their population, but most importantly, their demographics must be fully accounted for. Neutrons in the core are born, scattered, leaked, absorbed, and rejuvenated in a fully accounted physics-based manner.

When a reactor is in steady state operation, the reactor has achieved a kind of self-sustaining chain reaction which is call *critical* in the language of reactor operations. Criticality is a characterization that refers to the self-sustaining nature of the chain reaction. If the reactor has achieved criticality and stays at that state (through appropriate control effectors) we have a machine that disassembles atoms to produce electron motion, or heat with which we can do many different things including producing molecules such as hydrogen and methane, desalinate water or provide energy for a Mars colony.

A model of neutron behaviors can give us an easy-to-understand mathematical description of the core. A key idea is that a full-account of what happens to neutrons and with neutrons is possible, easy to formulate and can be tested with experiment. Since neutrons exist in large numbers, we use probabilities formulations even though we may refer to variables deterministically, as in variables of mechanical systems such as bridges and structures. For instance, we use a probability measure to model the fraction of the neutrons that may be lost by leaking out of the core; some may be absorbed unproductively (they produce no new neutrons), or absorbed in various components and structures inside the core; some may be absorbed in the fuel productively, that is disassembling fissile material; and, some absorptions may take place in material which we can refer to as "junk," again, without producing new neutrons.

A model of what happens to a neutron in the core is shown diagrammatically in Figure 3.6. A neutron born in fission (in the

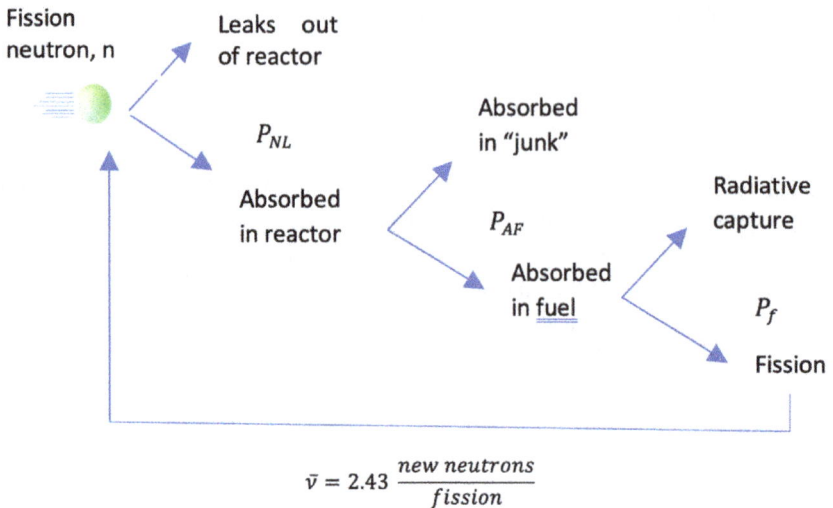

$$\bar{\nu} = 2.43 \; \frac{new \; neutrons}{fission}$$

Figure 3.6. Simple model of the nuclear chain reaction in a reactor. The neutron regeneration factor $\bar{\nu}$ equals 2.43 neutrons per fission in ^{235}U but 2.9 in the fission of ^{239}Pu. The cycle, known as the "neutron regeneration time," is approximately 1 ms in a PWR.

upper-left side of the diagram) proceeds through a life cycle, called "a generation," that in LWRs lasts approximately 1 μs. In fast reactors there is no moderator and the time of a generation is roughly one tenth of that, i.e., 0.1 μs. We think of the reactor generation as the "clock of the core," that is, a characteristic clock concept analogous to a "computer clock."

As discussed earlier, when we talk about "a neutron" we talk about a statistical representative of a very large population of neutrons. On average, the entire neutron population is renewed approximately every 1 μs. Every renewal is thought of as a *generation*. In steady state, the chain reaction is in "a critical state," that is, the total number of neutrons in one generation equals the total number of neutrons in the preceding generation. To capture this in a quantitative manner we introduced the concept of the *neutron multiplication factor*, k, defined as follows,

$$k \equiv \frac{\text{Total number of neutrons in a generation}}{\text{Total number of neutrons in the preceding generation}} \quad (3.5)$$

where

$k = 1$, The reactor power is in steady state; the reactor is "critical"
$k < 1$, The reactor power is decreasing; the reactor is "subcritical"
$k > 1$, The reactor power is increasing; the reactor is "supercritical"

Thus, k is an index of merit that tells us about the state of the reactor at any given time. Our goal is to determine k using the model displayed in Figure 3.6.

First, let's define the following probabilities.

$P_{\mathrm{NL}} \equiv$ Probability that a neutron does not leak out of the system
$P_{\mathrm{AF}} \equiv$ Conditional probability that if a neutron is absorbed, it is absorbed in the fuel
$P_f \equiv$ Conditional probability that if a neutron is absorbed in the fuel, it will induce fission

We recall that $\bar{\nu}$ is the number of neutrons per fission. Let's assume that we start with 1 neutron in the beginning. Then,

$$k = \bar{\nu}\, P_{\mathrm{NL}}\, P_{\mathrm{AF}}\, P_f \tag{3.6}$$

We are going to calculate first the two terms, P_{AF} and P_f. The term P_{NL} is a little more complicated as it requires establishing the space boundaries of the core from which neutrons can leak out and no longer participate in the nuclear chain reaction. Besides the concept of microscopic cross section, we need the concept of *neutron flux* and the concept of *macroscopic cross-section*.

What happens – probabilistically – when a neutron interacts not with one but many nuclei? The answer depends on the density of the nuclei. We define the macroscopic cross section, Σ, as follows,

$$\Sigma \equiv \sigma N \tag{3.7}$$

where

σ is the microscopic cross section,
N is the density of nuclei in units of $\left[\frac{\text{Number of nuclei}}{\text{Unit volume}}\right]$

The units of macroscopic cross-sections Σ are $\left[\frac{1}{cm}\right]$ and hence the interpretation of Σ is

$\Sigma \equiv$ Probability per cm of travel for a neutron to interact

Let us next define the important concept of neutron flux. If v is the speed of a neutron, we have

$$v \equiv \frac{\text{Distance traveled by neutron}}{\text{Unit time}} \tag{3.8}$$

The probability that the neutron interacts per unit time, gives the interaction frequency, F, in other words the frequency with which interactions occur, and is defined as follows:

$$F \equiv v\Sigma = \frac{\text{Probability that a neutron interacts}}{\text{Unit time}} \tag{3.9}$$

Consider a neutron density $n(\boldsymbol{r}, t)$ at some time t and a location inside the core described by the vector $\boldsymbol{r}$. To speak of a neutron density at a point $\boldsymbol{r}$, we mean that at time t there is an infinitesimally small volume $n(\boldsymbol{r}, t)d^3\boldsymbol{r}$ about the point $\boldsymbol{r}$ which contains $n(\boldsymbol{r})$ neutrons. The units of neutron density are $\left[\frac{\text{Number of neutrons}}{\text{Unit volume}}\right]$. From this and Eq. (3.9), we infer that number of interactions that this neutron engages in per unit time and per unit volume is,

$$n(\boldsymbol{r}, t)v\Sigma = \frac{\frac{\text{Number of interactions}}{\text{Unit time}}}{\text{Unit volume}} \tag{3.10}$$

The first two terms of Eq. (3.10) are very often found together. They are given a special name, *neutron flux*, denoted by the Greek letter $\Phi(\boldsymbol{r}, t)$, which is a function of space and time, or,

$$\Phi(\boldsymbol{r}, t) \equiv n(\boldsymbol{r}, t)v \tag{3.11}$$

Hence, Eq. (3.10) can be written as

$$n(\boldsymbol{r}, t)v\Sigma = \Sigma\Phi(\boldsymbol{r}, t) \tag{3.12}$$

Neutron flux $\Phi(\boldsymbol{r}, t)$ is an important variable in modeling the nuclear chain reaction. For convenience we will omit the spatial and time dependence and just use Φ unless it is necessary to explicitly indicate that the neutron flux is a scalar function of vector position and time. The neutron flux has units of intensity, that is

$$\left[\frac{\text{Number of neutrons}}{\text{Unit volume}} \times \frac{\text{Unit distance}}{\text{Unit Time}}\right] = \left[\frac{\text{Number of neutrons}}{(\text{Unit area}) \times (\text{Unit time})}\right]$$

For evaluating the neutron multiplication factor k let's translate into quantitative forms with the help of the neutron flux and macroscopic cross section the verbal definitions of P_{AF} and P_f. Recall,

$P_{\text{AF}} \equiv$ Conditional probability that if a neutron is absorbed, it is absorbed in the fuel

$P_f \equiv$ Conditional probability that if a neutron is absorbed in the fuel, it will induce fission

If we are considering a homogeneous core of volume V at time t, we can write the above as,

$$P_{\text{AF}} = \frac{\text{Number of neutrons absorbed in fuel}}{\text{Number of neutrons absorbed}},$$

$$\text{or,} \quad P_{\text{AF}} = \frac{\Phi \Sigma_a^F V t}{\Phi \Sigma_a V t} = \frac{\Sigma_a^F}{\Sigma_a} \tag{3.13}$$

where Σ_a^F is the macroscopic absorption cross section for absorptions in the fuel, and Σ_a is the total macroscopic absorption cross section.

The ratio $\frac{\Sigma_a^F}{\Sigma_a}$ in Eq. (3.13) is an important dimensionless parameter which can be seen as a figure of merit for how well the fissile material (fuel) is utilized. It is called the *fuel utilization factor f* and is defined as,

$$f \equiv \frac{\Sigma_a^F}{\Sigma_a} \tag{3.14}$$

For a heterogeneous core, or a quasi-homogeneous core where absorptions take place in moderator and structural material including the clad or tubing ensconcing the fuel pellets, it is rather easy to show that the fuel utilization factor f becomes,

$$f \equiv \frac{\Sigma_a^F}{\Sigma_a^F + \Sigma_a^M + \Sigma_a^C} \tag{3.15}$$

where Σ_a^F is the macroscopic cross section for neutrons absorbed in the fuel, Σ_a^M is the macroscopic cross section for absorption in the moderator, and Σ_a^C is the macroscopic cross section for absorption in the clad.

As far as the conditional probability that if a neutron is absorbed in the fuel, this absorption will lead to fission, we can quantify as follows,

$$P_f = \frac{\text{Number of thermal neutrons inducing fission}}{\text{Number of thermal neutrons absorbed in fuel}}$$

$$P_f = \frac{\Phi \Sigma_f^F V^F t}{\Phi \Sigma_a^F V^F t} = \frac{\Sigma_f^F}{\Sigma_a^F} = \frac{N \sigma_f^F}{N \sigma_a^F} = \frac{\sigma_f^F}{\sigma_a^F} \tag{3.16}$$

where N is the density of nuclei in the fuel, and we replaced the macroscopic cross section with microscopic cross sections in accordance with Eq. (3.7).

The ratio of these microscopic cross sections when multiplied by the (average) number of neutrons per fission, gives us a measure of the reproductive efficacy (in terms of neutrons) of the chain reaction, it is called the *neutron reproduction factor*, η, and is defined as follows,

$$\eta \equiv \bar{\nu} \, \frac{\sigma_f^F}{\sigma_a^F} \tag{3.17}$$

The *neutron reproduction factor*, η, gives us the average number of neutrons produced for more fissions, when one neutron is absorbed by the fissile material (fuel).

Using (3.13) and (3.17), the neutron multiplication factor of Eq. (3.6) can be written as

$$k = \eta f \, P_{\text{NL}} \tag{3.18}$$

The non-leakage probability term, P_{NL}, depends entirely on the geometry of the core. If the core is "infinite" in size (in nuclear terms) then $P_{\text{NL}} = 1$, that is the probability that no neutrons will leak is absolutely certain. Hence, (3.18) becomes,

$$k = \eta f \equiv k_\infty \tag{3.19}$$

This is the neutron multiplication for the asymptotic geometry of an infinite medium. Hence, we now have that the neutron multiplication factor is k_∞, multiplied by P_{NL}, or

$$k = k_\infty P_{\text{NL}} \tag{3.20}$$

To calculate the non-leakage probability in Eq. (3.20), we begin with a neutron accounting approach where all neutrons in a core volume V surrounded by a surface area A must be completely accounted for, that is,

$$
\left\{
\begin{array}{c}
\text{The rate of} \\
\text{change in the} \\
\text{number of} \\
\text{neutrons} \\
\text{in volume } V
\end{array}
\right\}
=
\left\{
\begin{array}{c}
\text{The rate of} \\
\text{production} \\
\text{of neutrons} \\
\text{in volume } V
\end{array}
\right\}
-
\left\{
\begin{array}{c}
\text{The rate of} \\
\text{absorption} \\
\text{of neutrons} \\
\text{in volume } V
\end{array}
\right\}
$$

$$
-
\left\{
\begin{array}{c}
\text{The rate of} \\
\text{leakage} \\
\text{of neutrons} \\
\text{from } A \text{ of } V
\end{array}
\right\}
\tag{3.21}
$$

This is called an "equation of continuity." It simply says that neutrons do not come from nowhere and do not disappear without reason. All terms in the curly brackets can be translated into mathematical expressions with the help of the variables of flux and macroscopic cross sections. For instance, the first term on the left is translated as follows,

$$
\left\{
\begin{array}{c}
\text{The rate of} \\
\text{change in the} \\
\text{number of} \\
\text{neutrons} \\
\text{in volume} V
\end{array}
\right\}
= \frac{\partial}{\partial t} \int_V n(\boldsymbol{r}, t) d^3 \boldsymbol{r}
\tag{3.22}
$$

where $n(\boldsymbol{r}, t)$ is the density of neutrons inside the reactor (core) volume.

The second term from the left in Eq. (3.21) is

$$\left\{ \begin{array}{c} \text{The rate of} \\ \text{production} \\ \text{of neutrons} \\ \text{in volume } V \end{array} \right\} = \int_V S(\boldsymbol{r}, t) d^3\boldsymbol{r} \qquad (3.23)$$

where $S(\boldsymbol{r}, t)$ is the fission neutron source term which accounts for new neutrons generated from the fission of fissile material inside the core volume V, that is,

$$S(\boldsymbol{r}, t) = \bar{\nu} \Sigma_f \Phi \qquad (3.24)$$

Hence, the rate of production of neutrons in V is

$$\left\{ \begin{array}{c} \text{The rate of} \\ \text{production} \\ \text{of neutrons} \\ \text{in volume } V \end{array} \right\} = \int_V \bar{\nu} \Sigma_f \Phi d^3\boldsymbol{r} \qquad (3.25)$$

The third term in Eq. (3.21) is,

$$\left\{ \begin{array}{c} \text{The rate of} \\ \text{absorption} \\ \text{of neutrons} \\ \text{in volume } V \end{array} \right\} = \int_V \Sigma_a \Phi d^3\boldsymbol{r} \qquad (3.26)$$

The fourth term calls for integrating over the surface A of the volume V, for which we use the neutron current $\boldsymbol{J}(\boldsymbol{rt})$, a vector variable of the flux of neutrons in a certain direction (along a unit vector). The number of neutrons streaming out of the surface A in the direction of a unit vector $\hat{u}$ perpendicular to the surface A is given by

$$\left\{ \begin{array}{c} \text{The rate of} \\ \text{leakage} \\ \text{of neutrons} \\ \text{from } A \text{ of } V \end{array} \right\} = \int_A \boldsymbol{J}(\boldsymbol{r}, t) \cdot \hat{u} d^2\boldsymbol{r} \qquad (3.27)$$

Equation (3.27) gives the leakage rate across the surface A through the surface integral on the right-hand side. It should be noted that the differential $d^2\boldsymbol{r}$ of the integral stands for an infinitesimal of surface area and the integration is over the boundary surface A of the core. This surface integral can be transformed into a volume integral in accordance the *divergence theorem* or *Gauss' theorem*, i.e.,

$$\int_A \boldsymbol{J}(\boldsymbol{r},t) \cdot \hat{u} \, d^2\boldsymbol{r} = \int_V \boldsymbol{\nabla} \cdot \boldsymbol{J}(\boldsymbol{r},t) d^3\boldsymbol{r} \qquad (3.28)$$

Hence, for the fourth term the following expression is obtained involving a volume integral,

$$\left\{ \begin{array}{c} \text{The rate of} \\ \text{leakage} \\ \text{of neutrons} \\ \text{from } A \text{ of } V \end{array} \right\} = \int_V \boldsymbol{\nabla} \cdot \boldsymbol{J}(\boldsymbol{r},t) d^3\boldsymbol{r} \qquad (3.29)$$

When we substitute these mathematical expressions in Eq. (3.21), we obtain the following equation,

$$\frac{\partial}{\partial t} \int_V n(\boldsymbol{r},t) d^3\boldsymbol{r} = \int_V \bar{\nu}\Sigma_f \Phi d^3\boldsymbol{r} - \int_V \Sigma_a \Phi d^3\boldsymbol{r} - \int_V \boldsymbol{\nabla} \cdot \boldsymbol{J}(\boldsymbol{r},t) d^3\boldsymbol{r} \qquad (3.30)$$

Since all terms involve the volume integrals over the volume V, Eq. (3.30) must be true at any time t and for any position $\boldsymbol{r}$ inside the volume V. Therefore, we can remove the volume dependence and write the continuity equation in its differential form, that is,

$$\frac{\partial}{\partial t} n(\boldsymbol{r},t) = \bar{\nu}\Sigma_f \Phi(\boldsymbol{r},t) - \Sigma_a \Phi(\boldsymbol{r},t) - \boldsymbol{\nabla} \cdot \boldsymbol{J}(\boldsymbol{r},t) \qquad (3.31)$$

Equation (3.31) is the *neutron transport equation* in differential form. Due to its foundational role in describing the conservation of neutrons within the reactor core, it is also called the *neutron continuity equation*.

It should be noted that Eq. (3.31) involves three different dependent variables, i.e., neutron flux, $\Phi(\boldsymbol{r},t)$, neutron density, $n(\boldsymbol{r},t)$, and neutron current, $\boldsymbol{J}(\boldsymbol{r},t)$. Neutron flux is the most important of the three. Hence, we express the other two in terms of neutron flux. First, we observe that collisions are the principal phenomenon that governs neutron movement. Hence, from diffusion theories of fluids we assume that the neutron current is proportional to the gradient of the flux, that is,

$$J(r, t) = -D\nabla\Phi(r, t) \tag{3.32}$$

where D is a diffusion coefficient (depends on the material in which collisions take place). Equation (3.32) is called *Fick's Law* and it is reasonably good approximation for large reactor cores (Lamarsh and Baratta, 2001). From Eq. (3.11), we have $\Phi(r, t) \equiv n(r, t)v$. Hence, the neutron transport equation can be written as,

$$\frac{1}{v}\frac{\partial}{\partial t}\Phi(r, t) = S(r, t) - \Sigma_a\Phi(r, t) + \nabla \cdot D\nabla\Phi(r, t) \tag{3.33}$$

Since the divergence of a divergence is the Laplacian operator, we have that,

$$\nabla \cdot D\nabla\Phi(r, t) = D\nabla^2\Phi(r, t) \tag{3.34}$$

Using Eq. (3.34), the neutron transport Eq. (3.33) can be rewritten as,

$$\frac{1}{v}\frac{\partial}{\partial t}\Phi(r, t) = S(r, t) - \Sigma_a\Phi(r, t) + D\nabla^2\Phi(r, t) \tag{3.35}$$

Fick's Law is applicable as said, hence Eq. (3.35) is known as *the time-dependent neutron diffusion equation.*

Equation (3.35) is an inhomogeneous, partial differential equation with initial and boundary conditions. Boundary conditions require that the neutron flux is zero at the surface A of the core volume V. The boundary condition is valid at a boundary slightly corrected for some transport phenomena not mentioned here for simplicity, the reader can find the relevant concepts at (Lamarsh and Baratta, 2001; Duderstadt and Hamilton, 1976; Tsoulfanidis, 1995). Associated with the equation are initial conditions,

$$\Phi(r, t = 0) = \Phi_0(r) \tag{3.36}$$

When the reactor is at steady state (typically 95% of the time, a power reactor is operating at steady state) the time-dependence of the flux can be eliminated. In addition, in many cases we can assume that neutrons see a relatively homogeneous mixture of atoms in their space. Under these assumptions, that is, steady-state, homogeneous and one-speed neutrons, Eq. (3.35) becomes,

$$D\nabla^2\Phi(\boldsymbol{r}) - \Sigma_a\Phi(\boldsymbol{r}) + S(\boldsymbol{r}) = 0 \tag{3.37}$$

The neutron source due to fission is

$$S(\boldsymbol{r}) = \bar{\nu}\,\Sigma_f\,\Phi(\boldsymbol{r}) \tag{3.38}$$

Inserting Eq. (3.38) into Eq. (3.37), we obtain

$$D\nabla^2\Phi(\boldsymbol{r}) - \Sigma_a\Phi(\boldsymbol{r}) + \bar{\nu}\,\Sigma_f\Phi(\boldsymbol{r}) = 0 \tag{3.39}$$

We use a concept called "thermal diffusion length L," which is a measure, in a root-mean-square sense, of the distance a neutron travels before absorption and is defined as

$$L^2 \equiv \frac{D}{\Sigma_a} \tag{3.40}$$

to rewrite Eq. (3.39) as

$$\nabla^2\Phi(\boldsymbol{r}) - \frac{\Sigma_a}{D}\Phi(\boldsymbol{r})\left[\bar{\nu}\frac{\Sigma_f}{\Sigma_a} - 1\right] = 0 \tag{3.41}$$

Since,

$$\bar{\nu}\frac{\Sigma_f}{\Sigma_a} = \bar{\nu}\frac{\Sigma_f}{\Sigma_a^F}\frac{\Sigma_a^F}{\Sigma_a} = \eta f \equiv k_\infty \tag{3.42}$$

We can rewrite Eq. (3.41) using (3.40) and (3.41) as

$$\nabla^2\Phi(\boldsymbol{r}) + \frac{1}{L^2}[k_\infty - 1]\Phi(\boldsymbol{r}) = 0 \tag{3.43}$$

The square of the term $\frac{1}{L^2}[k_\infty - 1]$ in Eq. (3.42) is defined as the "material buckling" B_m, or

$$B_m^2 \equiv \frac{1}{L^2}[k_\infty - 1] \tag{3.44}$$

Hence, the mathematical problem to solve is succinctly defined as an eigenvalue problem, that is,

$$\nabla^2\Phi(\boldsymbol{r}) + B_m^2\Phi(\boldsymbol{r}) = 0 \tag{3.45}$$

together with appropriate boundary condition (BC) which in this case is the requirement that the flux goes to zero a little distance away from the the surface of the core. This is a small distance d away from the actual surface, which is called "the extrapolated boundary" of the reactor core, the location of which is signified by vector, $\tilde{r}_s$. Hence, the BC is

$$\Phi(\tilde{r}_s) = 0 \qquad (3.46)$$

where $\tilde{r}_s = r_s + d$, with r_s stands for the vector description of the surface of the core and d can be estimated based on neutron characteristics and the material composition of the core.

The solution of Eq. (3.45) is known as an eigenvalue problem, meaning that only certain B's will satisfy the boundary condition (3.46). These B's are called the eigenvalues of the problem and they are a class of values, the smallest of which is called "Geometric Buckling" and is denoted as B_g. The name comes from the fact that it is a measure of the curvature of the flux. For a critical reactor, the smallest eigenvalue of (3.45) which we call geometric buckling has to equal the material buckling. This is another name for the critical-ity condition, that is, $k = 1$. The criticality condition can be thus rephrased as

$$B_m^2 = B_g^2 \qquad (3.47)$$

The non-leakage probability P_{NL} is defined as

$$P_{NL} \equiv \frac{\text{(Rate of neutron absorbion)}}{\text{(Rate of neutron absorbion)} + \text{(Rate of neutron leakage)}}$$

$$(3.48)$$

where

$$\text{Rate of neutron absorbion} = \int_V \Sigma_a \Phi(r) d^3r \qquad (3.49)$$

and, as before, using the Divergence Theorem and Fick's Law, we write

$$\text{Rate of neutron leakage} = \int_V D\nabla^2 \Phi(r, t) d^3r \qquad (3.50)$$

Plugging (3.49) and (3.50) into Eq. (3.48), we have

$$P_{\mathrm{NL}} \equiv \frac{\int_V \Sigma_a \Phi(\boldsymbol{r}) d^3 \boldsymbol{r}}{\int_V \Sigma_a \Phi(\boldsymbol{r}) d^3 \boldsymbol{r} + \int_V D \nabla^2 \Phi(\boldsymbol{r}) d^3 \boldsymbol{r}} \tag{3.51}$$

From Eq. (3.46) and assuming we have criticality (i.e., $B_m^2 = B_g^2$) we can write the second term in the numerator of (3.51) as

$$P_{\mathrm{NL}} \equiv \frac{\int_V \Sigma_a \Phi(\boldsymbol{r}) d^3 \boldsymbol{r}}{\int_V \Sigma_a \Phi(\boldsymbol{r}) d^3 \boldsymbol{r} + \int_V D B_g^2 \Phi(\boldsymbol{r}) d^3 \boldsymbol{r}} \tag{3.52}$$

$$P_{\mathrm{NL}} = \frac{\Sigma_a \int_V \Phi(\boldsymbol{r}) d^3 \boldsymbol{r}}{\Sigma_a \int_V \Phi(\boldsymbol{r}) d^3 \boldsymbol{r} + D B_g^2 \int_V \Phi(\boldsymbol{r}) d^3 \boldsymbol{r}}$$

$$= \frac{\Sigma_a \int_V \Phi(\boldsymbol{r}) d^3 \boldsymbol{r}}{(\Sigma_a + D B_g^2) \int_V \Phi(\boldsymbol{r}) d^3 \boldsymbol{r}}, \tag{3.53}$$

Dividing numerator and denominator by the volume integral in (3.53), we obtain,

$$P_{\mathrm{NL}} = \frac{\Sigma_a}{\Sigma_a + D B_g^2} \tag{3.54}$$

Recalling the definition of the thermal diffusion length as given by Eq. (3.40), that is, $L^2 \equiv \frac{D}{\Sigma_a}$, Eq. (3.54) becomes

$$P_{\mathrm{NL}} = \frac{1}{1 + L^2 B_g^2} \tag{3.55}$$

Equation (3.55) is saying that the probability of non-leakage depends on the geometry of the reactor, as embedded in the Geometric Buckling, which is the solution of a geometric problem. Different geometries give different geometric buckling. For instance, if the reactor shape is cylindrical with radius R and height H, the geometric buckling is the solution of Eq. (3.45) with boundary conditions given by (3.46) is as follows (Lamarsh and Baratta, 2001)

$$B_g^2 = B_1^2 = \left(\frac{\pi}{\widetilde{H}}\right)^2 + \left(\frac{\nu_0}{\widetilde{R}}\right)^2 \tag{3.56}$$

And the flux is

$$\Phi(r, z) = A J_0 \left(\frac{\nu_0}{\tilde{\tilde{H}}} r \right) \cos \left(\frac{\pi}{\tilde{\tilde{H}}} z \right) \tag{3.57}$$

where J_0 is the zeroth order Bessel function and ν_0 is its first zero, that is ν_0 is to J_0 like $\frac{\pi}{2}$ is to the cosine function (Boyce and DiPrima, 2022). It has a numerical value $\nu_0 = 2.45$.

Again, the criticality condition is as before, i.e.,

$$B_m^2 = B_g^2$$

From more rigorous neutron transport theory, see (Lamarsh and Baratta, 2001; Degerstedt, 1975), the diffusion coefficient in Eq. (3.37) is,

$$D = \frac{1}{3\Sigma_{\text{tr}}} = \frac{1}{3} \lambda_{\text{tr}} \tag{3.58}$$

where, Σ_{tr} is the macroscopic transport cross-section, which is the probability per unit length of neutron travel that a neutron will scatter with nuclei, and λ_{tr} is the mean free path (or average distance) that a neutron will travel before engaging in a transport-type interaction.

To add more sophistication to our model we have to distinguished between neutrons with thermal energies to the left of Figure 3.3 and fast energies to the right. In addition, we may add even more complexity modeling what happens in the epithermal region. Resonances in this region can absorb neutrons without leading to fission and the neutron absorption capacity of resonances may vary with macroscopic conditions in the reactor.

We recall that the effective multiplication factor as defined in Eq. (3.6) via the introduction of cross-sections and neutron flux. If only thermal neutrons are considered, the neutron multiplication factor is transformed into Eq. (3.18), i.e.,

$$k = \eta f \, P_{\text{NL}}$$

where η is the neutron regeneration factor, that is the number of neutrons produced per absorption of thermal neutron in the fuel (a fuel property); f is the probability that if a thermal neutron is absorbed, it is absorbed in the fuel; and, P_{NL} is the probability that a neutron will not leak out of the systems before absorption.

To account for fast neutrons, first we recognize that some neutrons can also be produced by the fission of fertile material like ^{238}U which is abundant within the fuel rods. We define a new parameter, ε, called the *fast fission factor*, as

$$\varepsilon \equiv \frac{\text{Total number of fission neutrons (thermal + fast fission)}}{\text{Number of neutrons from thermal fission}}$$

We can define p the resonance escape probability as,

p = Prob. that neutron goes from fast to thermal with no absorbion.

Thus, the neutron regeneration factor becomes,

$$\eta = \bar{\nu}\,\varepsilon \tag{3.59}$$

Additionally, the total *non-leakage probability* P_{NL} can be broken up in two parts, P_{FNL} or *fast non-leakage probability*, that is, the probability that a fast neutron will not leak out of the reactor (as fast neutron); and P_{TNL} or *thermal-non leakage probability*, that is, the probability that a thermal neutron will not leak out of the reactor. Hence the total non-leakage probability becomes

$$P_{NL} = P_{FNL} \times P_{TNL} \tag{3.60}$$

Expanding (3.18) using the above distinctions between fast and thermal neutrons we obtain the equation

$$k = P_{FNL}\, p\, P_{TNL}\, f\eta\varepsilon \tag{3.61}$$

Equation (3.61) is known as the *Six Factor Formula* due to the six product factors involved. If we isolate four of them in what is called the *Four Factor Formula*, that is,

$$k_\infty = pf\eta\varepsilon \tag{3.62}$$

The *six-factor formula*, Eq. (3.62), is written as

$$k = k_\infty\, P_{FNL}\, P_{TNL} \tag{3.63}$$

where, k_∞ as given by the *four-factor formula*, Eq. (3.62) is dependent entirely on the physical properties of the fuel.

For a typical PWR we have the following values for the parameters of the six-factor formula:

$$P_{FNL} = 0.97$$

$$p = 0.87$$

$$P_{TNL} = 0.99$$

$$\eta = 1.65 \tag{3.64}$$

$$\varepsilon = 1.02$$

$$f = 0.21$$

In the chain reaction, neutrons are born by fissile material, ^{235}U, ^{239}Pu, or ^{233}U, as fast neutrons, that is with kinetic energies of approximately 2 MeV. After birth, they slow down to become thermal neutrons with kinetic energies of approximately 0.025 eV. For neutrons to thermalize, they have to pass through the epithermal region (see Figure 3.3) also called the resonance region, where in the range of energies form 1 eV to 1 keV they may suffer (n, γ) reactions in resonances of fertile material, that is ^{238}U and ^{233}Th, and their contribution to the chain reaction is negligible.

Example 3.1. Calculate the optimal reactor shape of a PWR.

In the cylindrical core of the PWR shown in Figure 3.5 what is the optimal reactor shape. Calculate the ratio of the radius to the height of the core.

Solution: The extrapolated volume $\tilde{V}$ of the PWR cylindrical core is,

$$\tilde{V} = \pi(\tilde{R})^2 \tilde{H} \tag{3.1.1}$$

We recall that we use extrapolated radius and height for the cylindrical volume (all quantities with a tilda above, signify extrapolated quantities). For the reactor to be in a critical state, from Eq. (3.56) we have that

$$B_m^2 = B_g^2 = B_z^2 + B_r^2$$

or,

$$B_m^2 = \left(\frac{\pi}{\tilde{H}}\right)^2 + \left(\frac{\nu_0}{\tilde{R}}\right)^2$$

From which we obtain that

$$\widetilde{R}^2 = \frac{\nu_0^2}{B_m^2 - \left(\frac{\pi}{\widetilde{H}}\right)^2} \tag{3.1.2}$$

Substituting (3.1.2) into (3.1.1), we have that the volume of the reactor becomes,

$$\widetilde{V} = \pi \left(\frac{\nu_0^2}{B_m^2 - \left(\frac{\pi}{\widetilde{H}}\right)^2}\right)^2 \widetilde{H} \tag{3.1.3}$$

To optimize the shape of the reactor, essentially means minimizing the volume as given by Eq. (3.1.3). To find the minimum volume, we take the derivative of (3.1.3) with respect to the extrapolated height and we set it equal to zero. From that, we solve for the extrapolated height, that is

$$\frac{d\widetilde{V}}{dH} = \frac{\pi \nu_0^2}{\left[B_m^2 - \left(\frac{\pi}{\widetilde{H}}\right)^2\right]^2} \left\{B_m^2 - \left(\frac{3\pi}{\widetilde{H}}\right)^2\right\} = 0 \tag{3.1.4}$$

For (3.1.4) to be true, the expression within the curly brackets must be equal to zero, hence, we obtain

$$B_m^2 - \left(\frac{3\pi}{\widetilde{H}}\right)^2 = 0 \Rightarrow B_m^2 = \left(\frac{3\pi}{\widetilde{H}}\right)^2$$

And,

$$\widetilde{H} = \sqrt{3}\frac{\pi}{B_m} \tag{3.1.5}$$

Inserting (3.1.5) into Eq. (3.1.2), we obtain,

$$\widetilde{R}^2 = \frac{\nu_0^2}{B_m^2 - \frac{B_m^2}{3}} = \nu_0^2 \frac{3}{2B_m^2} \tag{3.1.6}$$

Using (3.1.5) and (3.1.6) and after taking the square root of (3.1.6), we form the ratio

$$\frac{\widetilde{R}}{\widetilde{H}} = \frac{\nu_0}{\sqrt{2\pi}} = \frac{2.45}{\sqrt{2\pi}} \approx 0.55 \tag{3.1.7}$$

Equation (3.1.7) says that we need minimum fuel for criticality in a PWR, if we make the diameter approximately equal to the height of the core, that is, if the core is arranged as a right cylinder.

In a typical PWR core we have approximately 200 assemblies of which approximately 60 are control assemblies while the remaining 140 are fuel assemblies. Each control assembly contains 20 control rods. Control assemblies is inserted in the core to absorb neutrons and thus control the fission rate. They are useful in startup maneuvers, maintaining power level, or shutting the reactor down.

In a fuel assembly, there are 15×15 fuel rods. Each fuel rod contains approximately 400 fuel pellets in a hermetically sealed zircaloy tube (Lamarsh and Baratta, 2001). Fuel pellets are made of UO_2 where the uranium initially contains up to 5% of the fissile isotope ^{235}U which is the actual nuclear fuel. The remaining 95% is ^{238}U which over the course of the operational life of the fuel, gets transmuted to the fissile ^{239}Pu which is also fuel.

Example 3.2. Calculate the radius of a PWR core for criticality.

Consider the cylindrical core of the PWR shown in Figure 3.5 and use the reactor model we have developed in order to determine its size. We assume that the coolant contains in addition to H_2O, 2210 parts per million (ppm) of boron, a natural shim, passively adjusting for reactivity changes due to the fissioning of fuel. The fuel is UO_2 enriched to 2.78 atoms per cent, 2.78 a/0, in ^{235}U. It is given that $\Sigma_{tr} = 3.62 \times 10^{-2} \frac{1}{cm}$.

Solution: To find the radius and the height of the cylindrical core, we make some simplifying assumptions regarding the neutron population, the principal driver of the chain reaction, and use thermohydraulic information which is obtained outside the neutron population and defines the capacity of the coolant flow to remove the heat produced by the chain reaction.

We will the steady state diffusion equation which for cylindrical coordinates gives us Eq. (3.57). First let us calculate the diffusion coefficient in terms of the macroscopic transport cross section from Eq. (3.58), that is,

$$D = \frac{1}{3\Sigma_{tr}}$$

For a large PWR, $\Sigma_{tr} = 3.62 \times 10^{-2}$ cm^{-1}. Thus, the diffusion coefficient is

$$D = \frac{1}{3(3.62 \times 10^{-2}\ \text{cm}^{-1})} = 9.21\ \text{cm} \qquad (3.2.1)$$

Assuming that the actual height of the core $H = 370$ cm is determined by thermal-hydraulic (heat removal) considerations, we will need to know the extrapolation distance, which will be added to the actual height (and radius),

$$d = 0.71\frac{1}{\Sigma_{tr}}$$
$$= 0.71\frac{1}{3.62 \times 10^{-2}\ \text{cm}^{-1}}$$
$$= 19.6\ \text{cm} \qquad (3.2.2)$$

Hence, the extrapolated height $\tilde{H} = H + 2d$, because the extrapolation distance d has to be at both bottom and top of the core. Thus,

$$\tilde{H} = H + 2d = 370\ \text{cm} + 2 \times 19.6\ \text{cm} = 409.2\ \text{cm}$$

Hence, the axial geometric buckling is

$$B_z^2 = \left(\frac{\pi}{\tilde{H}}\right)^2$$
$$= \left(\frac{\pi}{370\ \text{cm} + 2 \times 19.6\ \text{cm}}\right)^2$$
$$= 6.0 \times 10^{-5}\ \text{cm}^{-2} \qquad (3.2.3)$$

The total geometric buckling is the sum of the axial and the radial geometric buckling, that is,

$$B_g^2 = B_z^2 + B_r^2 \qquad (3.2.4)$$

When at steady state, or criticality, the total geometric buckling has to equal the material buckling, or

$$B_g^2 = B_z^2 + B_r^2 = B_m^2 \qquad (3.2.5)$$

To calculate the material buckling, we need to calculate k_∞

$$k_\infty = \bar{\nu}\frac{\Sigma_f}{\Sigma_a} = 2.43\frac{0.065 \text{ cm}^{-1}}{0.1532 \text{ cm}^{-1}} = 1.03 \qquad (3.2.6)$$

Thus, the material buckling is

$$B_m^2 \equiv \frac{1}{L^2}[k_\infty - 1] = \frac{\bar{\nu}\Sigma_f - \Sigma_a}{D}$$

or,

$$B_m^2 = \frac{0.157 \text{ cm}^{-1} - 0.1532 \text{ cm}^{-1}}{9.21 \text{ cm}} = 4.13 \times 10^{-4} \text{ cm}^{-2} \qquad (3.2.7)$$

From Eq. (3.2.5), we have,

$$B_r^2 = B_m^2 - B_z^2$$

or,

$$B_r^2 = B_m^2 - B_z^2$$
$$= 4.13 \times 10^{-4} \text{ cm}^{-2} - 6.0 \times 10^{-5} \text{ cm}^{-2}$$
$$= 3.53 \times 10^{-4} \text{ cm}^{-2}$$

But $B_r^2 = \left(\frac{\nu_0}{\tilde{R}}\right)^2$.

Where, the first number where the Bessel function of the first kind goes to zero is (Boyce and DiPrima, 2022),

$$\nu_0 = 2.45$$

Solving, we have

$$\tilde{R} = \frac{2.45}{\sqrt{3.53 \times 10^{-4} \text{ cm}^{-2}}} = 128 \text{ cm}$$

From the extrapolated radius, we have

$$\tilde{R} = R + d \text{ and rearranging, we get}$$
$$R = \tilde{R} - d = 128 \text{ cm} - 19.6 \text{ cm}$$

Hence, $R = 108.4$ cm is the radius required for the core to be critical in this type of reactor.

This is underestimating the radius of a typical PWR which is closer to $R = 180$ cm. This is so because we used one-speed model for the neutron population. In more advanced approaches we should have multi-speed neutron populations for the analysis. Also, the neutron regeneration factor $\bar{\nu}$ equals 2.43 neutrons per fission in ^{235}U but 2.9 in the fission of ^{239}Pu. The entire regeneration cycle, known as the "neutron regeneration time," is approximately 1 microsecond in a PWR. The neutron population dynamics, sometimes referred to as the "neutron economy," is all that is important for the chain reaction at any given time. Without sufficient neutrons, of the right energy and at the right places, there is no serious possibility to start and maintain a chain reaction.

3.3   Advanced Reactors

Machines that breed new fuel are fast reactors, where fast refers to the speed of neutrons. As shown in Figure 3.3, fast, epithermal and thermal are three categories of neutrons having corresponding kinetic energies ranging from the MeV to the eV range.

In fast reactors, neutrons are more useful when their energies are in the fast category, that is before they are moderated to the epithermal or thermal categories. Again, the epithermal range is fraught with difficulties because of the many resonances involved and the dynamic nature of resonances varying with temperature and influenced by core composition and size/shape.

Within the core, fuel occupies only about 30% of the volume, with the remaining space filled by coolant (approximately 50%) and structural materials (around 20%). While a spherical core would minimize neutron leakage, practical considerations favor a more manageable right-circular cylinder shape. Factors like heat transfer and fluid flow limitations constrain the core's height and here too as in Example 3.1 (which was for thermal reactors) a flatter shape limits leakage. Deviation from a spherical geometry result in increased neutron leakage, requiring a larger critical mass and consequently reducing the breeding ratio.

In addition to neutron speed categorizations, fast vs thermal, there are some key differences between fast and thermal reactors stemming primarily from their distinct neutron energy spectra.

First, fast reactors utilize highly enriched fuel (around 20% fissile material) compared to the low enrichment levels (0.7–3%) found in thermal reactors (Judd, 2014).

Second, the different fuel enrichment, combined with the absence of a neutron moderator, results in significantly smaller and more compact fast reactor cores (on the order of 1 m) compared to the larger cores of thermal reactors (around 3 m for light water reactors as seen in Example 3.2). This also leads to much higher power densities in fast reactors.

Third, the dominance of fast neutrons in fast reactors minimizes the impact of materials that strongly absorb thermal neutrons, such as xenon-135, samarium-147, and boron, which play a crucial role in thermal reactors. This reduces the significance of neutron poisoning effects and simplifies reactor control in fast reactors (Lamarsh and Baratta, 2001).

Fourth, the longer mean free path of fast neutrons leads to a more uniform neutron flux distribution within the fast reactor core, reducing the need for complex spatial calculations that are essential for analyzing neutron flux variations in thermal reactors. Fifth, while both reactor types exhibit temperature coefficients of reactivity, the underlying mechanisms differ. Fast reactors primarily rely on the Doppler effect and coolant expansion, while thermal reactors are more significantly influenced by moderator temperature changes.

Despite these simplifications, neutron flux calculations in fast reactors are more complex due to the wider range of neutron energies. This necessitates the use of multi-group calculations, unlike the simpler one-group or few-group models that are employed in thermal reactor analysis.

The neutron density within a reactor at steady state is a function of position, neutron energy, and the direction of neutron travel. This can be mathematically represented as

$$n = n(\boldsymbol{r}, E, \boldsymbol{\Omega}) \qquad (3.65)$$

where $n = n(\boldsymbol{r}, E, \boldsymbol{\Omega})$ = neutron density at position $\boldsymbol{r}$, with energy E, and traveling in direction $\boldsymbol{\Omega}$.

If the spatial and angular dependence of the neutron density is removed, we can show from the neutron transport equation that for a homogeneous reactor at criticality, that is when $k = 1$, or equivalently the material buckling equals the geometric buckling, $B_m^2 = B_g^2$, the equation for neutron density is reduced to

$$\frac{d}{dt}n(t) = \frac{k-1}{l}n(t) \qquad (3.66)$$

where

$k = \bar{\nu}\frac{\Sigma_f}{\Sigma_a}P_{NL}$

$l \equiv \frac{1}{v\Sigma_a}P_{NL}$, this is the mean neutron lifetime in a finite reactor

$$P_{NL} = \frac{1}{1+L^2 B_g^2}$$

The solution of (3.65) with initial condition $n(t = 0) \equiv n_0$ is

$$n(t) = n_0 e^{\frac{t}{T}} \qquad (3.67)$$

where $T \equiv \frac{l}{k-1}$ is the reactor period, that is the time it takes for the power (which is proportional to the neutron density) to grow by the exponential factor $e = 2.71828$.

For instance, if $l = 10^{-4}$ s and $k = 1.0025$, the reactor period is

$$T \equiv \frac{l}{k-1} = \frac{10^{-4}\text{ sec}}{1.0025 - 1} = 0.04\text{ s} \qquad (3.68)$$

This implies that the reactor power will increase by a factor of $e = 2.71828$ in 0.04 s. This is an intolerably fast change, making it very difficult to control the reactor. Luckily, there is an additional group of neutrons produced by fission products called "delayed neutrons." Such neutrons are produced by fission fragments called *delayed neutron precursors*, e.g., ^{87}Br. They are fission fragments whose β decay yields a daughter nucleus which subsequently decays yielding a delayed neutron, with some significant time delay. With delayed neutrons taken into account, the mean neutron lifetime above, is stretched from $l = 10^{-4}$ s to $l = 0.094$ s. Hence, the reactor period of (3.68) becomes

$$T \equiv \frac{l}{k-1} = \frac{0.094\text{ s}}{1.0025 - 1} = 37.5\text{ s} \qquad (3.69)$$

Comparing the (3.68) with (3.69), a major difference in reactor period is observed. With delayed neutrons, it is certain that the reactor is controllable (e.g., the reactor period is within the means of effectors such as control rods, to respond in a timely fashion and maintain criticality).

The neutron density obeys a linear version of the Boltzmann transport equation. The detailed form and derivation of this equation can be found in standard works on neutron transport (e.g., Ott and Bezella, 1989; Duderstadt and Hamilton, 1976). While the neutron transport equation is fundamental, its complexity makes it challenging to solve directly. The difficulty arises from two general problems: (a) The neutron density depends on six independent scalar variables (three spatial coordinates, two angular coordinates defining the direction of travel, and energy). (b) The nuclear cross-sections, which govern fission, scattering, and absorption, exhibit intricate variations with neutron energy.

To simplify the analysis, the "diffusion theory" approximation is often employed. This approach involves integrating the transport equation over all directions, leading to an energy-dependent diffusion equation. The diffusion equation is typically expressed in terms of the neutron flux, as shown in Eq. (3.35). The diffusion equation provides an accurate approximation under the assumption of isotropic or nearly isotropic neutron flux. This assumption generally holds true except in strongly absorbing media or regions where the properties of the medium change significantly over distances comparable to the mean free path of the neutrons.

In fast reactors, the long mean free paths of fast neutrons (typically 10 cm or more) and the relatively large dimensions of fuel elements and structural components (several millimeters) allow for the application of diffusion theory across significant portions of the reactor core. This is because, over distances comparable to the mean free path, the reactor core can be reasonably approximated as a homogeneous medium.

Diffusion theory also remains applicable for most practical purposes even within control rods, despite their neutron-absorbing properties. However, diffusion theory may introduce inaccuracies at the boundaries between different materials within the reactor core, such as the interface between the core and the breeder blanket. Exceptions where more advanced treatments are needed, include, but are not limited to, small fast reactors, heterogenous core configurations and neutron shielding calculations. In very small reactors, where the core dimensions are comparable to the neutron mean free path, diffusion theory may not accurately capture neutron behavior. Similarly, experimental reactors, such as the Purdue University Reactor 1 (PUR-1) utilize highly heterogeneous core configurations, such as plate-type assemblies with dimensions comparable to the mean free path and thus more advanced transport theory leads to more accurate analysis. Finally, some modified versions of diffusion theory may be required for accurate neutron shielding calculations.

An additional impediment to solve the diffusion equation in the above listed cases is the computational costs rises exponentially because we have to divide the continuous neutron energy spectrum is typically divided into a series of discrete energy groups. The energy range of interest (from approximately 10 MeV down to 0.1 eV or lower) is divided into distinct energy intervals: $E_1, E_2, \ldots, E_g$, where E_1 is the highest energy and E_g is the lowest. Hence, we need to have a neutron flux within each energy group and a separate diffusion equation can then be written for each energy group, accounting for neutron interactions (fission, scattering, absorption) within that specific energy category.

This multigroup approach provides a practical framework for solving the neutron transport equation in fast reactors and Eq. (3.37) for the different energy groups becomes,

$$-D_g\nabla^2\phi_g + \Sigma_{rg}\phi_g = \sum_{g'=1}^{g'=g}\Sigma_{sg'\to g}\phi_{g'} + \frac{1}{k}\chi_g\sum_{g'=1}^{g'=g}\bar{v}_{g'}\Sigma_{fg'}\phi_{g'} \quad (3.70)$$

This is not the most general form of the multigroup diffusion equations but it is the form usually encountered in fast reactor calculations.

The terms in the equations can be understood best by thinking of them as a balance of neutrons. If each term is multiplied by a small volume dV, Eq. (3.70) is giving us a full accounting of balance between the rates at which group g neutrons appear in and disappear from dV.

The first term in the above equation is the net rate at which neutrons diffuse out of dV. D_g is a diffusion coefficient, and is assumed to be independent of position within dV.

The second term is the rate at which neutrons at the small volume dV are lost from group g. Σ_{rg} is the "group removal cross-section."

It includes the effects of absorption and both inelastic and elastic scattering. Scattering interactions which do not remove the neutron from the group (known as "in-group scattering") do not appear in Σ_{rg}.

The third term is the rate at which neutrons are scattered into group $g \cdot \Sigma_{sg' \to g}$ is the transfer cross-section from group g' to g. It includes both inelastic and elastic scattering, but unless the groups are narrow or hydrogen is present elastic scattering can take a neutron from one group to the next but not further. This elastic scattering appears usually in the $\Sigma_{s(g-1) \to g}$ cross-sections, while the $\Sigma_{sg' \to g}$ for $g' < g-1$ involve only inelastic scattering. In Eq. (3.70), the third term goes only up to $g - 1$ because it can be assumed that there is no up-scattering. This is not always true for a thermal reactor.

The fourth term in the diffusion equation represents the rate at which neutrons are born into energy group "g" through fission, where again, $\bar{\nu}_{g'}$ is the average number of neutrons emitted per fission event, which depends on the energy of the neutron inducing fission. The macroscopic fission cross-section for neutrons in energy group "g–1" at position "r" is represented by the term, $\Sigma_{fg'} \, \phi_{g'}$. The fraction of fission neutrons born into energy group "g" is represented by χ_g, which is a factor that ensures that the neutron distribution among energy groups after fission accurately reflects the energy spectrum of fission neutrons.

Before solving the multigroup diffusion Eq. (3.70) it is necessary to determine the group constants (such as $\Sigma_{f,g-1}$, χ_g, etc.) for each energy group. These constants are derived from the energy-dependent microscopic cross-sections of the materials present in the reactor.

This involves averaging the microscopic cross-sections over the energy range of each group. The specific averaging scheme used can significantly influence the accuracy of the multigroup calculations.

In principle, achieving high accuracy in multigroup diffusion calculations would require a very large number of energy groups, potentially thousands, to adequately capture the energy-dependent variations of nuclear cross-sections. However, practical limitations arise because of the computational cost and data requirement. The first refers to solving the diffusion equations for a large number of energy groups. Especially when considering spatial variations across the reactor core, we have extremely demanding calculations, potentially exceeding feasible computational resources. The data requirement refers to cross-section data for each energy group. This significantly increases the data requirements and complexity of the calculations. Therefore, practical considerations necessitate the use of a more manageable number of energy groups, typically around 20–30.

A key challenge in solving the neutron transport equation lies in estimating the spatial variation of the neutron flux within each energy group. To address this, a simplifying assumption is made: the spatial dependence of the neutron flux is assumed to be separable from the energy dependence. This is achieved by introducing the concept of "buckling," a constant parameter (B^2) that characterizes the spatial variation of the neutron flux. As seen in Eq. (3.45), buckling is a measure of the curvature of the neutron flux within the reactor core. By assuming that the spatial dependence of the flux in all energy groups can be represented by the same buckling factor, the problem becomes significantly simpler.

In addition, in order to accurately capture the energy dependence of neutron interactions, the energy range is divided into a large number of "fine groups" (denoted by subscript "n"). Within each fine group, the cross-sections are assumed to be relatively constant.

Finally, if the buckling, B^2, is chosen correctly to represent a critical reactor, the following condition holds

$$\phi_n = \sum_{n'=1}^{n-1} \frac{\Sigma_{sn' \to n}\phi_{n'} + \chi_n}{D_g B^2 + \Sigma_{rn}} \qquad (3.71)$$

Equation (3.71) describes the spatial variation of the neutron flux within each fine group, assuming a simple exponential decay.

The fine group fluxes can be normalized such that

$$\sum_{n=1}^{N} \nu_n \Sigma_{fn} \phi_n = 1 \qquad (3.72)$$

The normalization of Eq. (3.72) ensures that the sum of the fluxes in all fine groups equals unity. By employing this simplified approach and utilizing the concept of buckling, it becomes possible to estimate the energy spectrum of the neutron flux, which can then be used to calculate the group constants for more accurate multigroup diffusion calculations.

3.4   Producing Energy and Breeding Fuel

The fissile isotopes ^{235}U, ^{239}Pu, and ^{233}U are the principal fuels of the Atomic energy transition. The fertile isotopes, that is the isotopes that can breed fissile isotopes, are ^{238}U and ^{233}Th (Thorium-233). Both are very abundant on Earth and can provide all of humanity's energy needs for tens of thousands of years. While other fissile and fertile isotopes exist, these two fertile and three fissile isotopes are the most important. In this section we examine how breeding of new fuel takes place in the Atomic energy transition.

When a neutron is captured by a nucleus of fertile ^{238}U, a rather short and quick chain of transmutations produces new fuel. This is the fissile isotope ^{239}Pu which is like the fissile ^{235}U, only better in terms of achieving and maintaining a chain reaction. This qualitative superiority of the fissile material ^{239}Pu compared to $^{235}_{92}$U is due to several factors, an important one being that the fission of ^{239}Pu produces more neutrons than the fission of ^{235}U.

Specifically, while the fission of ^{239}Pu gives on average 2.9 neutrons per fission, the fission of ^{235}U gives 2.4 neutrons per fission. Hence the overall neutron economy inside a fast reactor involves a greater number of neutrons. To see the importance of this small numerical difference, 2.9 vs 2.4, let us consider what happens to neutrons produced in fission, as shown in Figure 3.1.

In the fission of $^{239}_{9}$Pu, of the 2.9 neutrons produced only 1 is needed to maintain the chain reaction. The rest 1.9 can be used for transforming ^{238}U into ^{239}Pu. A fraction of 1.9 can be lost into other absorptions which do not lead to breeding (or fission) but still 1.9 is a lot better to start with as compared with 1.4 neutrons. In other words, half a neutron can make a big difference, in a pool of nuclei available for either fission or breeding interactions.

The process can go on *ad infinitum* with ^{239}Pu producing energy and more ^{239}Pu fuel at the same time. The nuclear transformations and the times involved are shown in the diagram below.

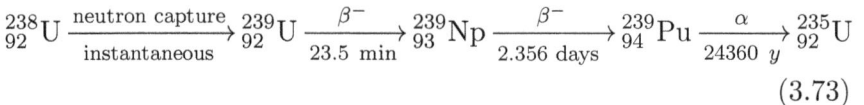

$$^{238}_{92}\text{U} \xrightarrow[\text{instantaneous}]{\text{neutron capture}} {}^{239}_{92}\text{U} \xrightarrow[\text{23.5 min}]{\beta^-} {}^{239}_{93}\text{Np} \xrightarrow[\text{2.356 days}]{\beta^-} {}^{239}_{94}\text{Pu} \xrightarrow[\text{24360 } y]{\alpha} {}^{235}_{92}\text{U}$$

$$(3.73)$$

Starting from the left side of the diagram, a $^{238}_{92}$U nucleus captures (absorbs) a neutron and nearly instantaneously becomes a nucleus

of $^{239}_{92}$U. In 23.5 minutes, this nucleus (statistically speaking) transforms or transmutes itself by turning a neutron into a proton (inside
the body of the nucleus) in a process called β^- decay since this transformation gives off an electron (a negatively charged particle) and a
particle called anti-neutrino. As a result, the atomic number of the
nucleus is now 93, hence we have the element $^{239}_{93}$Np formed which
in 2.356 days, again through β^- decays into $^{239}_{94}$Pu. Strictly speaking the times shown in the diagram represent the half-lives of the
radioactive decay, that is the time required for half of the nuclei of a
certain isotope to transform themselves into the new identity resulting from the radioactive decay. For practical purposes, given that the
operational life of a nuclear reactor is in the order of 100 years, when
compared to the long half-life $^{239}_{94}$Pu, i.e.,

$$100 \text{ years} \ll 24360 \text{ years}$$

Hence, we can consider $^{239}_{94}$Pu to be the final product of this decay
chain. This is the new fissile material, the new fuel. It all starts with
the fertile $^{238}_{92}$U, an isotope from which we can produce new fuel.

However, if $^{239}_{94}$Pu is not used as fuel, it is a radioactive material
that produces a radioactivity called α *radiation* for at least 24,360
years. This presents a formidable problem for nuclear waste management because the length of time is significantly beyond the predictive
horizon of the scientific method that forms the basis for all engineering design. Simply said, there is no engineering system that can be
designed to isolate $^{239}_{94}$Pu from the environment for 24,360 years. The
call to do so (the Nuclear Waste Act) is a political stratagem that
likely prioritized non-proliferation to energy production.

Instead of trying to isolate plutonium from the environment for
tenths of thousands of years, $^{239}_{94}$Pu should be burned as fuel. Fast
reactors can be thus seen as incinerators of $^{239}_{94}$Pu and of other isotopes of plutonium. And, again, burning $^{239}_{94}$Pu can produce energy as
well as new fuel. Otherwise, there is a nuclear waste problem which
is a meta-physical problem, not a problem that has a solution, as it
requires custodianship of a substance for at least 24,360 years.

It should be noted that $^{239}_{94}$Pu exhibits nuclear properties very
similar to the other fissile material, $^{235}_{92}$U, including the ability to
undergo fission when bombarded with neutrons of any energy as
seen in Figure 3.2. But, it all starts with $^{238}_{92}$U, the "fertile" isotope
of uranium.

Like the fertile $^{238}_{92}$U, there is a fertile isotope from the thorium family of isotopes that is, $^{233}_{90}$Th. This is the only naturally occurring isotope of thorium. Its behavior closely resembles that of $^{238}_{92}$U. When $^{233}_{90}$Th captures a neutron, it undergoes a series of radioactive decays as show in the following equation.

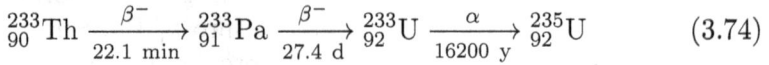

$$^{233}_{90}\text{Th} \xrightarrow[\text{22.1 min}]{\beta^-} \,^{233}_{91}\text{Pa} \xrightarrow[\text{27.4 d}]{\beta^-} \,^{233}_{92}\text{U} \xrightarrow[\text{16200 y}]{\alpha} \,^{235}_{92}\text{U} \qquad (3.74)$$

Thus, we have two naturally occurring fertile isotopes: $^{238}_{92}$U and $^{233}_{90}$Th from which we can breed new fuel, that is, produce fissile isotopes.

The average number of neutrons released per fission is denoted by "nu" ($\bar{\nu}$). The bar above nu indicates average. It varies depending on the isotope and the kinetic energy of the incoming neutron, but typically averages around 2.5.

A chain reaction can only occur when the value of $\bar{\nu}$ is greater than 1. However, for a breeder reactor, $\bar{\nu}$ must be significantly greater than 1. This ensures that, while one neutron sustains the chain reaction, additional neutrons are captured by fertile material, leading to the production of more fissile material than is consumed.

This unique and remarkable aspect of nuclear power suggests that properly designed reactors can generate substantial energy while simultaneously producing more fuel than they consume. A breeder reactor can generate more fuel than it consumes in the sense that it produces more fissile material than it consumes, if it is continuously supplied with fertile material, which it turns into fuel.

As shown in Figure 3.3, resonances can absorb neutrons unproductively. To know if breeding is possible, we use the average number of neutrons generated per neutron absorbed, a quantity denoted by η and called, the "thermal fission factor," which is given below

$$\eta \equiv \bar{\nu} \frac{\sigma_f}{(\sigma_f + \sigma_c)}. \qquad (3.75)$$

The "thermal fission factor," is a measure of how feasible is breeding. It represents the average number of neutrons produced per fission, adjusted by a factor less than one to account for neutron losses due to various capture events. The thermal fission factor, η, is a function of neutron energy, and its variation with energy is shown in Figure 3.2 for the two fissile isotopes ^{235}U and ^{239}Pu and the fertile ^{239}U.

Of the neutrons produced, only one is required to sustain the chain reaction. The remaining neutrons can be lost through diffusion out of the reactor or capture by other materials (like coolant or structural components). However, some neutrons can be captured by fertile nuclei, leading to the production of new fissile material.

If the parameter L denotes the number of neutrons lost per neutron absorbed in fissile material while the number captured in fertile material is denoted by C, we observe that C represents the number of fissile nuclei produced per fissile nucleus destroyed. The number of neutrons capture in fertile material can be approximated as

$$C \approx \eta - 1 - L \qquad (3.76)$$

The parameter C is indeed an approximation, that is, it doesn't account for all possible neutron interactions. If C is greater than one, indicating that more fissile material is produced than is consumed, it is referred to as the *breeding ratio*. If C is less than one, it is termed the *conversion ratio*. The neutron losses, L, (cannot be reduced below approximately 0.2. Consequently, breeding is only feasible if η (the number of neutrons produced per neutron absorbed) exceeds 2.2.

The population of neutrons in a nuclear reactor core, where a self-sustained chain reaction is domiciled, may be affected by the resonances shown in Figure 3.3. Resonance effects are important in all cores but even more important in fast reactors breeding new fuel. Many nuclei, particularly heavy elements like uranium and plutonium, exhibit sharp resonances in their cross-sections over specific energy ranges (typically between 1 eV and 100 eV) as shown in Figure 3.3. These resonances significantly influence neutron interactions and must be accurately accounted for in reactor modeling. As we see in Section 3.5, to accurately model these resonances requires slicing the neutron energy space in a very large number of energy intervals, also called "groups," to capture the detailed variations in cross-sections. The number of energy groups can be very large (in the thousands), presenting us with major computation challenges.

Near an isolated resonance peak, the microscopic cross-section for a specific neutron reaction, for instance "neutron capture," can be approximated by the following expression (Judd, 2014),

$$\sigma_c = \sigma_{co} \frac{\left(\frac{E_o}{E_c}\right)^{\frac{1}{2}}}{1 + \frac{4(E_c - E_o)^2}{\Gamma^2}} \tag{3.77}$$

where

σ_{co} = capture cross-section at E_o,
E_o = energy of rh resonance peak,
Γ = the "width" of the resonance (full width half-max),
E_c = the kinetic energy of the neutron and nucleus relative to the center of mass of the neutron-nucleus system.

If the nucleus is stationary, we have that

$$E_c = \frac{E}{1 + \frac{1}{A}} \tag{3.78}$$

where

A = the ratio of the mass of the nucleus to that of the neutron, and,
E = the neutron energy.

A nucleus can never really be stationary (this is a modeling idealization, which is true only in absolute zero, on the Kelvin scale). It actually moves at random due to thermal agitation and the effect on the cross-section can be estimated if the probability distribution of velocities of the nucleus is known. As in many thermal modeling situations, we assume, that the nucleus velocities have a Maxwell–Boltzmann distribution. The resulting distribution of the component of velocity parallel to a certain direction (which we can take as the direction of the neutron direction) can be estimated for atoms of $^{238}_{92}\text{U}$ at various temperatures (Judd, 2014).

From this the probability distribution of E_c can be deduced, and hence the mean cross-section for a neutron of energy E. This is given by

$$\bar{\sigma}_c(E, T) \cong \sigma_{co} \sqrt{\left(\frac{E_o}{E}\right)} \psi(\zeta, x) \tag{3.79}$$

where $k =$ Boltzmann's constant

$$\psi(\zeta, x) \equiv \frac{\zeta}{2\sqrt{\pi}} \int_{-\infty}^{+\infty} \left[\frac{e^{-\frac{(x-y)^2 \zeta^2}{4}}}{(1+y)^2} \right] dy$$

$$\zeta^2 \equiv \Gamma^2 \frac{A}{4E_o kT}$$

$$x \equiv \frac{2(E - E_o)}{\Gamma}$$

The approximations used in the Eq. (3.65) result in negligible errors in most fast reactor applications.

The impact of thermal motion of nuclei on neutron interactions can be significant. Even at room temperature, nuclei within the reactor fuel are in constant motion due to thermal agitation. This thermal motion influences the apparent neutron cross-section. If a nucleus is moving towards an approaching neutron, the relative velocity between the neutron and the nucleus increases, effectively increasing the neutron energy from the neutron's perspective. Conversely, if the nucleus is moving away from the neutron, the relative velocity decreases, and the apparent neutron energy is reduced. This effect, known as "Doppler Broadening," effectively broadens the observed width of the resonance. On the other hand, the probability distribution of the nuclear velocities (determined by temperature) can be used to calculate the effect of thermal motion on the apparent neutron cross-section.

Thermal agitation of nuclei within the fuel broadens the observed width of nuclear resonances, influencing the overall neutron interaction probabilities and consequently affecting reactor behavior. This Doppler broadening effect plays a crucial role in reactor safety and stability.

Fast reactors, utilizing any of the three fissile materials, can achieve breeding, with Pu-239 offering the most significant margin. However, breeding with U-235 in a fast reactor becomes challenging if neutron energies fall substantially below 1 MeV. In all cases, higher neutron energies generally result in better breeding ratios.

Thermal reactors utilizing ^{235}U may be capable of marginal breeding, but the margin is very slim. Other thermal reactors cannot breed but can still convert a considerable amount of fertile material into fissile material. Conversion ratios for ^{235}U-fueled thermal reactors range from 0.6 (for Light Water Reactors) to 0.8 (for heavy-water and gas-cooled reactors). Neutron absorption by hydrogen contributes significantly to the relatively high value of L in light water reactors.

Currently, the most favored breeder system utilizes a combination of ^{238}U and ^{239}Pu in fast reactors. Several experimental and prototype reactors have been built, and commercial plants are currently under investigation (especially in a smaller size SMRs like the Xe-100, a High Temperature Gas Cooled Reactor which reaches temperatures of 750°C to produce Hydrogen).

While thermal breeder reactors utilizing ^{233}Th and ^{233}U have been considered as an alternative, they present significant challenges where AI may have a significant role in addressing them.

Example 3.3. In an infinitely large reactor (i.e., the size of reactor is such that the escape of neutrons from the reactor is negligible), with no sources or sinks, all neutrons are thermalized, and the speed of neutrons obeys a Maxwellian distribution, i.e.,

$$n(v)dv = n_0 \frac{4\pi}{\left(2\pi\frac{kT}{m}\right)^{\frac{3}{2}}} v^2 e^{-\frac{m}{2}\frac{v^2}{kT}} dv \qquad (3.3.1)$$

where, $n(v)$ = number of neutrons per unit volume with speeds in dv about v. Assume n_0 is the total number of neutrons per unit volume. Find, (a) the most probable speed of the neutrons at thermal equilibrium under Maxwellian speed distribution, and (b) the kinetic energy of the neutrons at the probable speed (in eV).

Solution: The total number of neutrons n_0 must be the sum or integral of all neutrons of all different speeds under a Maxwellian distribution.

Hence, according to Eq. (3.3.1) the total number of neutrons per unit volume is

$$n_0 = \int_{v=0}^{v\to\infty} n(v)dv$$

The units for n_0 are $\left[\frac{\text{number of neutrons}}{\text{unit volume}}\right]$. In Eq. (3.3.1), we have that,

$$k = 1.38 \times 10^{-23} \frac{J}{{}^\circ K} \equiv \text{Boltzmann Constant},$$

$$T = \text{Temperature in degrees Kelvin}, {}^\circ K$$

$$m = \text{mass of neutron}$$

The Maxwellian distribution does not describe exactly the distribution of neutrons as a function of speed, but for the purpose of solving the problem, we accept the modeling assumption, which is often made, that the neutron distribution in the reactor is Maxwellian.

(a) Let's call the most probable speed of neutrons, v_p. Then, it is easy to see that the most probable speed is given by the derivative of the neutron distribution, evaluated at that speed, that is,

$$\left.\frac{dn(v)}{dv}\right|_{v=v_p} = 0 \tag{3.3.2}$$

Substituting (3.3.1) into (3.3.2) and evaluating the resulting expression at $v = v_p$ we obtain:

$$v_p = \sqrt{\frac{2kT}{m}} \tag{3.3.3}$$

Let's assume that we are at room temperature (the thermalization temperature), that is, $T = 20^\circ C$ which in the Kelvin scale is equal to $T = (20 + 273)^\circ K = 293^\circ K$ t.

Substituting the temperature in degrees Kelvin, Boltzmann's constant, and the mass of the neutron in Eq. (3.3.3) we obtain that the most probable speed of the neutrons at room temperature is

$$v_p = 2200 \; \frac{\text{m}}{\text{sec}}$$

Because of this result, thermal microscopic cross sections are usually given at $2200 \frac{\text{m}}{\text{sec}}$.

(b) The kinetic energy of neutrons having the most probable speed is,

$$KE = \frac{1}{2}mv_p^2 = kT = 0.025 \text{ eV}$$

Hence, the 2200 $\frac{m}{sec}$ cross sections are equivalent to 0.025 eV thermal cross section.

Example 3.4. What is the mean free path for a thermal neutron (KE = 0.025 eV) in graphite. The microscopic scattering cross section of neutrons interacting with carbon is $\sigma_s = 4.8$ bn and the microscopic absorption cross section is $\sigma_a = 4.0 \times 10^{-3}$ bn We recall that carbon $^{12}_{6}C$ is a bad neutron absorber because it is "doubly magic" that is it has $Z = 6$ and $A = 12$ which are both "magic numbers." The density of graphite is $\rho = 1.6 \frac{g}{cm^3}$. Recall that one mole of graphite equals 12.01 grams (g) and that Avogadro's number is $N_A = 6.023 \times 10^{23} \frac{particles}{mole}$.

Solution: Let us first calculate the atomic density N of graphite i

$$N = \left[1.6 \, \frac{g}{cm^3}\right] \times \left[\frac{1 \text{ mole}}{12.01 \text{ g}}\right] \times \left[\frac{N_A}{1 \text{ mole}}\right]$$

or,

$$N = \left[1.6 \, \frac{g}{cm^3}\right] \times \left[\frac{1 \text{ mole}}{12.01 \text{ g}}\right] \times \left[\frac{6.023 \times 10^{23} \, \frac{atoms}{mole}}{1 \text{ mole}}\right]$$

Hence, the atomic density of graphite is

$$N = 0.0802 \times 10^{24} \, \frac{atoms}{cm^3}$$

To calculate the macroscopic scattering cross section Σ_s and the macroscopic absorption cross section Σ_a we multiply the atomic density with the corresponding microscopic cross sections σ_s and σ_a respectively, i.e.,

$$\Sigma_s = N\sigma_s = 0.385 \, \frac{1}{cm}$$

and,

$$\Sigma_a = N\sigma_a = 3.2 \times 10^{-4} \; \frac{1}{\text{cm}}$$

The total macroscopic cross section is the sum of macroscopic absorption and scattering cross sections, that is,

$$\Sigma_{\text{total}} = \Sigma_s + \Sigma_a \approx 0.385 \; \text{cm}^{-1}$$

The mean free path is the inverse of the total macroscopic cross section, that is

$$mfp = \frac{1}{\Sigma_{\text{total}}} = \frac{1}{0.385 \; \text{cm}^{-1}} = 2.6 \; \text{cm}$$

Hence, a neutron travels a distance of approximately 2.6 cm before interacting with a carbon atom.

3.5 Fueling the Atomic Energy Transition

To fuel the Atomic energy transition we use a framework collectively referred as the *nuclear fuel cycle*, which involves various activities including, but not limited to, the mining of uranium, breeding new fuel, recycling of used fuel as an option for nuclear energy and safe disposal of any remaining nuclear waste. The "front end" of the nuclear fuel cycle starts with uranium mining and milling, conversion, enrichment and fuel fabrication. On the other hand, the "back end" begins after uranium has spent several years in a reactor. After it is removed, as used fuel may undergo a further series of steps including temporary storage, reprocessing, and recycling before the waste produced is disposed. The entire process is schematically illustrated in Figure 3.7, where in various stages the principal chemical forms or the fuel are identified.

The primary motivation for developing fast breeder reactors lies in their potential to significantly enhance the utilization of available uranium (and thorium) resources by thousands, possibly hundreds of thousands of years (OECD/NEA, 2020).

Consider a uranium-fueled reactor where N atoms of ^{235}U undergoing fission. Concurrently, $C \times N$ new fissile atoms (^{239}Pu) are

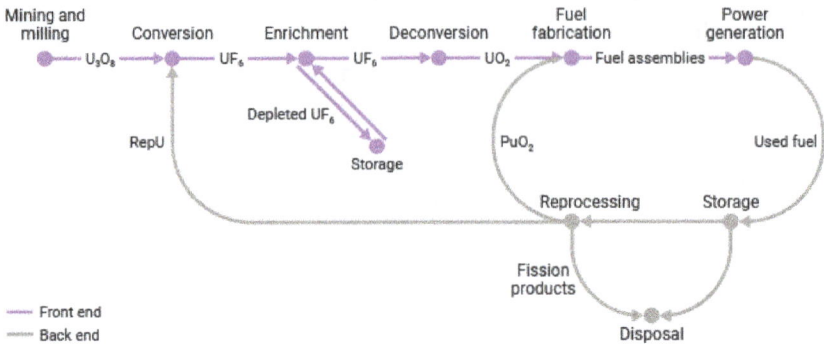

Figure 3.7. The stages of the nuclear fuel cycle (WNA, 2024).

produced through neutron capture and subsequent decay of ^{238}U as shown in the diagram of Eq. (3.1).

If these newly produced fissile atoms are subsequently fissioned within the same reactor, assuming a constant conversion ratio (C), another N fissile atoms will be generated. This process can theoretically continue indefinitely.

Therefore, the total number of atoms fissioned in this scenario is given by the following sequence

$$N + C \times N + C \times (C \times N) + C \times (C \times (C \times N)) + \cdots \quad (3.80)$$

The above sequence demonstrates the potential for a breeder reactor to significantly amplify the amount of energy extracted from a given quantity of fuel through the production of new fuel.

It should be noted that this a simplification for illustrative purposes. In actuality, the conversion ratio (C) may not remain constant but rather vary over time (mostly due to changes in the fuel composition). However, the core principle remains: a breeder reactor can potentially utilize a much larger fraction of the available uranium or thorium resources compared to conventional reactors.

$$N(1 + C + C^2 + C^3 + \cdots) \quad (3.81)$$

If $C < 1$, the series converges and its sum is

$$N(1 + C + C^2 + C^3 + \cdots) = N\left(\frac{1}{1-C}\right) \quad (3.82)$$

If the fuel is natural uranium, the new fissile isotopes N cannot exceed 0.7% of the total number of uranium atoms supplied. If the reactor is a thermal reactor with a conversion ratio of 0.7 and the plutonium bred is recycled indefinitely, the total number of atoms fissioned is approximately

$$\left(\frac{0.7}{1 - 0.7} \right) \approx 2.4\%$$

of the number of uranium atoms supplied (Judd, 2014).

Even in an idealized scenario, not all the theoretically fissionable atoms can be utilized in a real-world reactor. During fuel processing to remove fission products and separate excess plutonium, some fissile material is inevitably lost. Additionally, some plutonium is converted into higher isotopes, further reducing the available amount. As a result, thermal reactors can effectively utilize only about 2% of the uranium present in natural uranium.

In contrast, when a breeder reactor is involved, for instance, with $C > 1$, the series of fissions theoretically diverges, allowing, in principle, the fission of all supplied fertile material. However, in practice, some losses are unavoidable, and a more practically achievable limit is approximately 50% utilization of the fertile feed.

Consequently, fast breeder reactors have the potential to fission approximately 25 times more atoms than thermal reactors from a given quantity of natural uranium, leading to the extraction of roughly 25 times more energy.

To assess the significance of this difference, it's crucial to understand the available quantities of uranium and thorium.

The amount of accessible uranium depends on the price we're willing to pay for extraction. At a price of $150 per kilogram, approximately 6.5 million tons of "yellowcake" (a uranium concentrate) are likely available. Fissioning 2% of this uranium would yield about 9×10^{21} Joules of heat, roughly equivalent to 10^{18} British Thermal Units or BTUs (defined as 100 Q = 1 Quadrillion BTUs). In comparison, the world's estimated oil reserves could potentially yield about 12,000 Q, while annual global energy consumption currently stands at around 600 Q (OECD/NEA, 2020).

Breeder reactors, however, could potentially extract around 1,200 Q from the same 6.5 million tons of yellowcake, significantly surpassing the energy yield from all the world's oil. This enhanced energy extraction has another crucial implication: it allows for a substantial increase in the price that can be justified for uranium ore. If the price were to rise to \$500 per kilogram, the available uranium reserves could potentially increase tenfold to a hundredfold, unlocking an additional 50,000 Q of energy. This potential surpasses even the estimated energy yield from all the world's coal reserves (approximately 30,000 Q). Furthermore, at even higher prices, it might become economically viable to extract uranium from seawater, which contains an estimated 10^9 Q of uranium.

While thermal nuclear reactors, particularly non-breeder types, can undoubtedly contribute to global energy needs, they cannot fully address long-term energy shortages due to the limited utilization of uranium resources. Breeder reactors, on the other hand, offer the potential to significantly expand the available energy from uranium, fundamentally altering the long-term energy outlook.

It is beyond the scope of the Atomic energy transition fundamentals, but generally relying heavily on breeder reactors may present potential challenges and risks. Alternative energy sources, such as nuclear fusion (e.g., deuterium-deuterium fusion with a potential yield of 19^{10} Q) and renewable sources (harnessing solar energy, which provides around 5,000 Q per year to Earth), also offer promising avenues for future energy needs. However, while fusion technology is still under development, and the large-scale utilization of renewable sources still faces significant challenges, breeder reactors are likely to play a crucial role in meeting our future energy demands.

Early fast reactor development (before the 1960s) prioritized achieving high breeding ratios. This necessitated minimizing neutron moderation, leading to the use of metal fuels (enriched uranium or plutonium alloys) within the reactor core. These metal fuels allowed for operation at higher temperatures.

Fast reactors around the world are counted in the fingers of one hand. They are very small in numbers. Yet, they are mostly developed through international collaboration with data being shared, and technical challenges addressed collectively. While collaboration facilitates knowledge sharing and prevents significant divergence in design approaches among different countries, it also inadvertently limits the exploration of alternative coolant options, such as helium, due to the widespread adoption of liquid metals.

The small critical masses of these early reactors necessitated high-density coolants to achieve sufficient power output. Hydrogenous coolants were avoided due to their moderating properties, leading to the adoption of liquid metals like sodium or sodium-potassium alloys. Mercury, while used in some early reactors, was later largely abandoned due to its toxicity, cost, and low boiling point.

Many neutrons leaked from these small cores, prompting the use of surrounding "blankets" of natural or depleted uranium to capture these neutrons and breed more fissile material.

The first generation of low-power experimental fast reactors (Clementine, EBR-1, BR-1, BR-2, ZEPHYR, ZEUS) were built in the late 1940s and early 1950s to demonstrate the feasibility of breeding and gather crucial nuclear data. While some utilized plutonium fuel (more readily available in the early years), these reactors had very small cores, with EBR-1 (a 1.2 MW reactor) being the largest at 6 liters in volume.

To achieve higher power levels, the core volume of these early reactors needed to be increased to maintain acceptable heat fluxes and accommodate increased coolant flow. This led to the development of larger reactors like EBR-II and EFFBR in the USA and DFR in the UK. Initially envisioned as prototypes for commercial power plants, these reactors ultimately became testbeds for the next generation of oxide fuels, signaling a shift away from metal fuels in fast reactor development.

The focus shifted beyond achieving a high breeding ratio around the 1960s. It became evident that a profitable fast reactor requires more than just breeding. The high cost of fissile material, fabrication of fuel elements, and subsequent reprocessing significantly impacts the overall economics.

Fuel cannot remain in the reactor indefinitely due to several factors. Firstly, the fissile material is gradually consumed during the fission process, although this is partially offset by the production of new fissile material through breeding. Secondly, the accumulation of fission products within the fuel disrupts its structure, causing swelling and potentially corroding the cladding. Additionally, the cladding material itself is weakened by neutron bombardment, increasing the risk of fuel leakage into the coolant.

Once the fuel has reached its irradiation limit, it must be removed from the reactor and stored to allow for the decay of highly radioactive fission products. Subsequently, the spent fuel undergoes chemical reprocessing to remove fission products and recover remaining fissile material. This recovered material is then refabricated into new fuel elements and returned to the reactor. The frequency of this fuel cycle significantly impacts the overall cost, as it necessitates frequent reprocessing and increases the time during which expensive fuel remains out of service.

The amount of irradiation a fuel can withstand before needing replacement is termed "burnup." It can be quantified either by the fraction of uranium and plutonium (or thorium and uranium in thorium-based cycles) atoms that undergo fission or by the total amount of heat generated by the fuel. These two measures are essentially equivalent, as each fission event releases a relatively consistent amount of energy, and the fuel mass is primarily comprised of heavy atoms (uranium, plutonium, or thorium). Theoretically, fissioning all heavy atoms in a ton of fuel would yield approximately 10^6 MW-days of energy.

Early experience with metal fuels revealed a limited burnup capacity of around 1% (or 10,000 MWd/ton). In contrast, oxide fuels (UO_2, PuO_2, or mixed oxides) demonstrated significantly higher burnup capabilities, reaching 10% or more. The high reprocessing costs associated with low burnup fuels made metal fuels economically unviable.

Furthermore, metal fuels exhibit limited high-temperature performance. Phase changes within the metal itself and compatibility issues with cladding materials at temperatures exceeding 250°C pose significant challenges. These limitations restrict the achievable thermodynamic efficiency and, consequently, the electrical power output. These factors ultimately led to the rejection of metal fuels in favor of oxide fuels for most fast reactor designs.

While oxide fuels offer significant advantages over metal fuels, they also present certain limitations. The presence of oxygen within the oxide fuel acts as a partial neutron moderator, reducing the average neutron energy and consequently decreasing the breeding ratio.

Additionally, the low thermal conductivity of oxide fuels necessitates the use of slender fuel elements, increasing manufacturing costs.

However, oxide fuels also possess several advantages. The lower neutron energy levels enhance the Doppler effect, a negative feedback mechanism that helps to stabilize reactor operation. This is particularly beneficial in large reactors where positive feedback effects can arise. Moreover, oxide fuels can operate at higher temperatures and can be clad in stainless steel, a more cost-effective material compared to the refractory metals required for metal fuels. Furthermore, extensive experience with oxide fuels in thermal reactors provides a strong foundation for their use in fast reactors.

While alternative fuels like carbides (UC and PuC) offer potential benefits such as higher thermal conductivity and density, their limited understanding and potential challenges (like corrosion and fission gas accumulation) have favored the continued use of oxide fuels for the time being.

Other fuel types, such as *cermets* (a mixture of stainless steel and oxide), have also been explored. However, the significant neutron absorption by the steel component in *cermets* significantly hinders the breeding ratio, limiting their viability.

The LAMPRE experiment investigated the use of molten plutonium fuel clad in tantalum. However, challenges related to corrosion and the accumulation of fission gas bubbles within the molten fuel led to the abandonment of this approach.

The combination of mixed oxide fuel, stainless steel cladding, and sodium coolant has emerged as the de facto standard for fast reactor development. This commonality in design choices has resulted in a remarkable degree of similarity among current and planned fast reactors like the 600 MWe BN-600 connected to the grid in Russia, the 500 MWe Prototype Fast Breeder Reactor (PFBR) near completion in India, or the operational 29 MWe China Experimental Fast Reactor (CEFR). China has under construction a 600 MWe fast reactor called the CFR-600 in Fujian province.

International collaboration has led to prioritizing carbide fuels or gas coolants, and the focus has been on oxide-fueled, sodium-cooled reactors, at least within the foreseeable future. Reactor physics plays a crucial role in reactor design. In addition, heat transfer, structural integrity, metallurgy, and safety play an important role with many design options resulting from compromises amongst various factors ensuring safety and economic viability.

3.6   Nuclear Fusion in the Atomic Energy Transition

Controlled thermonuclear fusion is an important technology of the Atomic energy transition. Although the technological progress of fusion has been slow, the promise remains great. Relatively abundant isotopes of hydrogen such as deuterium are its basic fuel. Such isotopes, when they overcome the force of electrostatic repulsion, they fuse together to form heavier elements and release significant quantities of energy. The inspiration for fusion is the Sun, where similar process produces the energy responsible for all life on Earth. Hydrogen in ordinary water, has an atom of deuterium for every 6,500 atoms of ordinary hydrogen. Given the vast amounts of water in our Blue Planet, if fusion technology becomes practical to deploy, it will remove energy as a serious constraint to human progress.

Atoms can overcome electrostatic repulsion either through firing lasers at targets (called "inertia confinement fusion"), or by squeezing them in strong magnetic fields (called "magnetic confinement fusion"). In addition to energy, copious quantities of neutrons of rather high kinetic energies are released in the process. Just like in fission, these neutrons can be used to convert fertile material to fissile fuel. Thus, here too, in addition to energy we can breed new fuel (Glasstone and Lovberg, 1975; Miyamoto, 1980).

Although fusion was demonstrated by the hydrogen bomb in 1951, controlled nuclear fusion remains largely in research and development, with promising breakthroughs appearing in recent years. The Lawrence Livermore National Laboratory announced that ignition was achieved in late 2022. Firing powerful lasers at a tiny deuterium-tritium capsule (the target) at the laboratory's National Ignition Facility (NIF), more energy was produced than the amount of energy of laser energy delivered at the target. This is a historic milestone in inertia confinement fusion. Significant progress is recently observed in magnetic confinement with new and more powerful magnets and AI tools to control the process in ways that were not possible past magnetic confinement machines.

The fundamental fusion reactions of interest are:

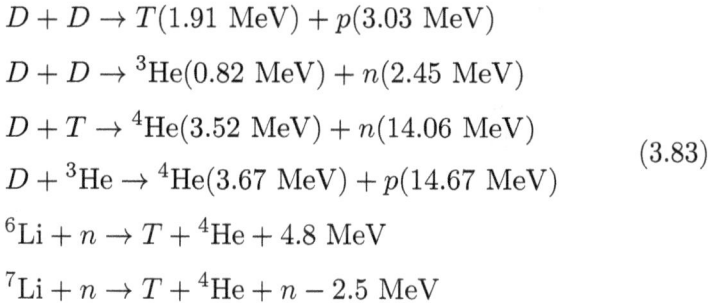

$$D + D \rightarrow T(1.91 \text{ MeV}) + p(3.03 \text{ MeV})$$

$$D + D \rightarrow {}^3\text{He}(0.82 \text{ MeV}) + n(2.45 \text{ MeV})$$

$$D + T \rightarrow {}^4\text{He}(3.52 \text{ MeV}) + n(14.06 \text{ MeV})$$

$$D + {}^3\text{He} \rightarrow {}^4\text{He}(3.67 \text{ MeV}) + p(14.67 \text{ MeV})$$

$$(3.83)$$

$$^6\text{Li} + n \rightarrow T + {}^4\text{He} + 4.8 \text{ MeV}$$

$$^7\text{Li} + n \rightarrow T + {}^4\text{He} + n - 2.5 \text{ MeV}$$

In (3.83) D stands for deuteron (the nuclei of deuterium, a hydrogen isotope that has a neutron and a proton in the nucleus and an electron in orbit), that is, ^2H; T stands for triton (the tritium isotope of hydrogen with two neutrons in the nucleus, ^3H; ^3He is the isotope of helium with one neutron in the nucleus (in addition to the two protons that define the element He); ^3He is the isotope with two neutrons in its nucleus; ^6Li is the isotope of lithium with three neutrons in the nucleus (in addition to the three protons that define the element); ^6Li is the isotope of lithium with four neutrons in its nucleus.

A figure of merit for the ease to release energy in the Atomic energy transition is the binding energy per nucleon (neutron or proton). The smallest the binding energy per nucleon the easier it is to release energy. The binding energy per nucleon is smaller in very light nuclei, such as the ones in (3.83) and also at very big nuclei such as the ones in (3.73) and (3.74). Hence, large quantities of energy can be released when the light nuclei of the isotopes of (3.83) are fused.

Although deuterium could alone be the fuel needed for fusion reactors as indicated in the (3.83), the electrostatic repulsion of D ions makes this a very difficult option in technological terms. On the other hand, the deuterium-tritium $(D + T)$ reaction is easier to achieve, although Tritium is rather rare in nature. It can be fabricated through the interaction of neutrons with lithium, as seen in the last two reactions of (3.83). Here too, breeding can be achieved in the sense that some initial quantity of tritium is used to achieve ignition, but after that, the required tritium can be regenerated from the neutrons produced in the fusion reactions (Glasstone and Lovberg, 1975). Again, we see that the unique feature of the Atomic energy transition, that is, fuel regeneration as seen in breeding fissile material, holds true for nuclear fusion as well (Delaporte-Mathurin *et al.*, 2025).

The nuclei of (3.71) when inside an operating fusion reactor poses astronomical temperatures exceeding 150 million degrees Kelvin. They exist in a plasma state (the fourth state of matter). Their densities are rather low, that is in the order of about 10^{15} nuclei/cm^3 but because of high temperatures their kinetic energies are very high.

For instance, a deuterium plasma with a particle density of 10^{15} deuterons/cm^3 and a temperature of 100 keV has deuterons with temperatures of 700 million degrees Kelvin and mean free path greater that 10^{10} cm. If the fusion reactor has a diameter of 10^2 cm, a deuteron nucleus will collide with the wall of the reactor 10^{10} times before interacting with another deuteron in a $D + D$ reaction. Each time it hits the wall a deuteron nucleus loses approximately 1 keV of energy. But when it hits and fuses with another D nucleus, the $D - D$ reaction releases approximately 10 MeV $= 10^4$ keV. Thus, the ratio of the energy loss to the energy gained is

$$\frac{10^8 \text{ keV}}{10^4 \text{ keV}} = 10^4 \qquad (3.84)$$

This is huge disadvantage as far as energy production goes, which implies that to reduce the numerator in (3.72), the plasma particles must not come in contact with a material wall (notwithstanding the huge damage to such a wall). The plasma must be confined by electric and magnetic fields, the former being possible but very difficult while the latter is the basis of almost all magnetic confinement reactors such as tokamaks, stellarators, pinch machines and mirror systems (Glasstone and Lovberg, 1975). An alternative to magnetic confinement is various accelerator approaches where beams of high energy particles collide with a solid or plasma target to achieve fusion, such as the NIF experiment mentioned earlier.

In addition to confinement, nuclear fusion reactors call for special material to withstand the harsh radiation environment including high energy protons and neutrons generated in (3.83). One measure of the durability and reliability of such materials is dpa or *displacements per atom*. It counts the number of atoms displaced from their original positions through collisions with high-energy neutrons. High dpa values signify significant radiation damage which produces detrimental effects on material containing the plasma, including, but not limited to embrittlement, swelling and structural degradation.

The $D+T$ reaction in (3.83) is of great interest as the reaction of choice of fusion reactors. However, this reaction produces high energy neutrons (14.1 MeV) which strike the reactor vessel and internal parts causing structural damage. Material that will withstand dpa greater than 150 are sought for commercial fusion reactors, a daunting challenge stimulating major quest to discover new materials.

Problems

Problem 3.1. The microscopic cross-sections for $^{235}_{92}$U at a neutron energy of 1 MeV are provided as follows:

- Scattering cross-section (σ_s) = 4 b
- Inelastic scattering cross-section (σ_i) = 1.4 b
- Fission cross-section (σ_f) = 1.2 b
- Absorption cross-section (σ_a) = 1.3 b

Assume that the cross-sections for neutron-emission and charged-particle reactions are negligible. Using this information, determine the following:

(a) The total microscopic cross-section (σ_T).
(b) The ratio of the capture cross-section to the fission cross-section, denoted as α.

Problem 3.2. Calculate the macroscopic cross section of $^{239}_{94}$Pu at thermal energy. The microscopic cross section of thermal neutrons interacting with plutonium-239 is $\sigma_s = 7.7$ b and the microscopic absorption cross section is $\sigma_a = 1011.3$ b. The nominal density of plutonium-239 is $\rho = 19.6$ g/cm^3.

Problem 3.3. In a bare sphere of $^{235}_{92}$U, 60% of the fission neutrons escape without causing further reactions. Given that the average neutron reproduction factor η for this system is 3.4, calculate the multiplication factor k of the sphere.

Problem 3.4. How much of the coal fuel with a heat content of 3×10^7 J/kg would need to be burned to release the same amount of energy as the fissioning of 10 g of ^{235}U? *Note*: Each fission of a ^{235}U nucleus releases approximately 200 MeV of energy, while 1 MeV = 1.602×10^{-13} J.

Problem 3.5. In a specific thermal reactor fueled with partially enriched uranium, it is observed that 11% of the fission neutrons are absorbed in the resonance region of ^{238}U, and 4% escape from the reactor during the slowing down process. Additionally, 6% of the neutrons that complete the slowing down phase subsequently escape from the reactor. Among the slow neutrons that remain in the reactor, 84% are absorbed by the fuel, with 76% of these being absorbed by ^{235}U. Assume that the reproduction factor is $\eta = 2.068$. Calculate the multiplication factor of this reactor.

Problem 3.6. A reactor is to be built with fuel rods of 1.6 cm in diameter, and a liquid moderator with a 1:1 1:2 volume ratio of moderator to fuel. What will the distance between nearest fuel centerlines be for a square lattice?

Problem 3.7. A thermal reactor fueled with uranium operates at 1.0 kW. The operator is to increase the power to 10 MW over a four-hour span of time. What reactor period should she put the reactor on?

Problem 3.8. A thermal reactor fueled with uranium operates at 100 W. The operators put in on a 30-minute period. How long will the reactor take to reach a power of 250 kW?

Problem 3.9. If a fusion reactor starts with 1 kg of deuterium (^{2}H) and 50% of the deuterium undergoes fusion to form tritium, how much tritium (^{3}H) in grams is produced? *Note*: In a fusion reactor, two deuterium nuclei can fuse together to create the following reaction:

$$^{2}H + {}^{2}H \rightarrow {}^{3}H + p$$

Problem 3.10. In a deuterium-deuterium (D–D) fusion reaction, one possible outcome is:

$$^{2}H + {}^{2}H \rightarrow {}^{3}H + p + 4.03 \text{ MeV}$$

If 1 gram of deuterium undergoes this fusion reaction, calculate the total energy released in joules. *Note*: 1 mole of deuterium weights 2 g/mol, and 1 MeV = 1.602×10^{-13} J.

References

Boyce, W.E. and DiPrima, R.C., *Elementary Differential Equations and Boundary Value Problems*, 12th edn., John Wiley & Sons, NY, 2022.

Delaporte-Mathurin, R., *et al.*, "Advancing tritium self-sufficiency in fusion power plants: Insights from the BABY experiment," *IAEA Nuclear Fusion*, 65, 1–11, 2025.

Duderstadt, J.J. and Hamilton, L.J., *Nuclear Reactor Analysis*, John Wiley & Sons, New York, NY, 1976.

Economou, E.N., *A Short Journey from Quarks to the Universe*, Springer, Heidelberg, 2011.

Glasstone, S. and Lovberg, R.H., *Controlled Thermonuclear Reactions: An Introduction to Theory and Experiment*, Robert E. Krieger Publishing Company, Huntington, NY, 1975.

Judd, A.M., *An Introduction to the Engineering of Fast Nuclear Reactors*, Cambridge University Press, New York, NY, 2014.

Lamarsh, J.R. and Baratta, A.J., *Introduction to Nuclear Engineering*, 3rd edn., Prentice Hall, Upper Saddle River, NJ, 2001.

Miyamoto, K., *Plasma Physics for Nuclear Fusion*, MIT Press, Boston, 1980.

OECD/NEA, IAEA, *Uranium 2020: Resources, Production and Demand*, Uranium Red Book, Paris, 2020.

Ott, K.O. and Bezella, W.A., *Introductory Nuclear Reactor Statics*, Revised edn., American Nuclear Society, Westmont, IL, 1989.

Prantikos, K., Chatzidakis, S., Tsoukalas, L.H., and Heifetz, A., "Physics-informed neural network with transfer learning (TL-PINN) based on domain similarity measure for prediction of nuclear reactor transients," *Nature Scientific Reports*, 13, 16840, 2023.

Prantikos, K., Pallikaras, V., Pantopoulou, M., Tsoukalas, L.H., and Heifetz, A., "Agentic retrieval augmented generation for advanced reactor thermal hydraulic system," *Proceedings of the 2024 American Nuclear Society Winter Meeting*, Orlando, FL, 2024.

Tsoulfanidis, N., *Measurement and Detection of Radiation*, 2nd edn., Taylor and Francis, London, 1995.

World Nuclear Association, Nuclear fuel cycle overview, https://world-nuclear.org/information-library/nuclear-fuel-cycle/introduction/nuclear-fuel-cycle-overview, Updated Monday, 2024.

Chapter 4

AI for Energy

4.1 AI Inferencing Support for Energy

Infrastructures are crucial for the functioning and progress of society. The energy infrastructure can justly be considered as the *infrastructure of infrastructures par excellence*. Pivoting on the energy infrastructure fuels all others including, health, shelter, industry, transportation, education and food production. The emerging AI infrastructure is a new pivot for all infrastructures, including energy. But AI needs lots of energy. In fact, the energy needs of AI infrastructure will be the main driver of the Atomic energy transition. Working together, AI and atomic energy, they can produce a secure, sustainable and prosperous future for Humanity.

AI needs atomic energy. At the same time AI may well be the philosopher's stone in enabling scientific and engineering discoveries that make atomic energy safer, cheaper and secure from non-proliferation, while at the same time through breeding its own fuel, it may free Humanity from energy as a fundamental barrier to human growth (see Chapter 3). Nuclear digitization and AI go hand to hand with new reactors producing energy and fuel at the same time. Conversely, nuclear reactors are an excellent long-term energy source for large scale deployment of the AI infrastructure providing inferencing support for it and all other infrastructures.

In this chapter we explore the strong linkages of AI and energy, focusing primarily on AI as it promises long-term gains in energy resilience, sustainability and efficiency. It must be emphasized that

AI goes beyond computing to provide inferencing support on a very large scale. Again though, automating inferencing requires gargantuan quantities of high-quality energy. Conversely, managing the complexity of the energy infrastructure requires immense decision support from AI and its adjoint, machine learning (ML).

Nonstop data gathering in increasingly digitized infrastructures far exceeds human capacities for data processing and analysis. This includes examining the data and correcting or rejecting data that are clearly erroneous. Generally, the available data are outputs of the systems. If the inputs are also available, it is theoretically possible, in the absence of noise, to obtain the system matrix model simply by dividing the output data matrix by the input data matrix. Usually, the experimenter is trying to obtain one or more critical parameters (time constants, sensitivity coefficients, etc.), and the evaluation of virtually all parameters is influenced by the presence of noise. If the inputs are not available, it is sometimes possible to use multiple outputs in a cross-correlation model to eliminate the need for inputs while at the same time of the influence of noise in the measurements or in the system.

AI uses a plethora of technologies synergistically which run on Graphical Processing Units (GPUs) and Tensor Processing Units (TPUs) enabled by ever more powerful hardware (e.g., a 2 nm chip packing 50 billion transistors in the area of a fingernail). For instance, bots running large language models (LLMs) making AI a household name, utilize a combination of graph theory, deep neural networks using groups of simple neural nets searching huge spaces structured as partially ordered sets (POSETS) with multiple interactions amongst themselves including learning from each other and supporting each other's functionality.[1] GPU and TPUs have enabled the rise of artificial intelligence (AI) and ML as a new infrastructure for inferencing support. But this is a two-way street. GPU and TPUs need a lot of energy to perform complex calculations needed to train AI models, while AI and ML and they inferencing support they provide will help us use energy efficiently and proceed towards the atomic energy.

---

[1]For an accessible introduction to language as computation the reader may consult Chapter 1 of book *Fuzzy Logic: Applications in Artificial Intelligence, Big Data, and* Machine Learning.

All AI technologies can be grouped in two categories. First, statistical AI which includes, but is not limited to, ML algorithms such as they abound in neural computing, genetic and evolutionary computing, decision trees, random forest, support vector machines and Bayesian approaches. Second, logical AI which includes, fuzzy logic, propositional logic, predicate logic, description logic, logic programming, abductive logic programming, situation calculus, temporal logic and modal logic. It should be noted that within these two categories there are several subcategorizations of methods and technologies. For instance, ML algorithms, we have a group of algorithms that belong to Supervised Learning, others that belong to Unsupervised Learning as well as separate groups of Probabilistic Models and Statistical Inference. Statistical AI is very powerful especially because it provides data-driven technologies of creating and utilizing models from the plethora of data available through digitization. Likewise, to delineate domains of knowledge, records, information and semantic nets, logical AI is crucial for decisions pertaining to the veracity and validity of ML (Tsoukalas, 2023; Tsoukalas and Uhrig, 1997).

Applications of AI for inferencing support fundamentally involves some variant of neural learning and logic, e.g., fuzzy logic. Evolutionary computing with genetic algorithms can provide additional growth and plasticity to models build with neural and logical approaches. These three, neural, fuzzy and genetic algorithms taken in isolation are by no means the most important approaches. But when used synergistically they help us understand how AI provide inferencing support in the energy infrastructure. The approaches have been developed and applied to engineering systems for decades. Neural networks, fuzzy logic, and genetic algorithms have shown to be of value in the surveillance, diagnostics and operation of industrial facilities. For instance, the surveillance and diagnostics of nuclear power plants (plants with digital instrumentation and control systems or I&Cs) can be performed remotely through Digital Twins using AI models built from data. Specific components such as check valves, instrumentation systems, and rotating machinery can be maintained through data-driven approaches, were data trains AM models, verifies their relevance and transfers knowledge across different areas. Sensor surveillance, calibration verification, diagnostics of both plant transients and specific faults, efficiency optimization, vibration analysis, loose parts monitoring, and adaptive and/or optimal control can all be achieved through AI. The synergistic benefits of combining the use of statistical AI techniques such as neural networks, with logical AI systems, such as fuzzy logic and genetic algorithms are illustrated in several application. AI may provide inferencing support for fossil-fired power plants, chemical process facilities, high performance aerospace systems, financial market issues and educational systems preparing technical workforce.

AI is all about inferencing support. As an infrastructure, it promises to exponentially increase our capacity to improve all other infrastructures, including the power grid which is posted for major growth to enable the electrification of transportation. Several AI technologies are highlighted here. They are expected to provide inferencing support in managing and controlling large, geographically dispersed and complex energy systems such as the grid, pipelines and energy storage. Each technology is sufficiently complex to warrant complete books as well as continuing technical journals sponsored by scientific and engineering societies (e.g., the series of Transactions published by IEEE in such areas as neural networks, evolutionary computing, fuzzy systems, etc.). It is our expectation that the reader will gain sufficient knowledge from these short descriptions of the different technologies that they can understand the AI inferencing support to the surveillance, diagnostics, prognostics and other applications of the energy infrastructure.

4.2   Neural Computing

A neural network, is a data processing system consisting of a large number of simple, highly interconnected processing elements in an architecture inspired by the structure of the cerebral cortex portion of the brain. Imitating Nature, a process called *biomimesis*, gives us confidence that computational tasks easily done by humans, may be replicated using neural networks. Hence, neural networks are often capable of doing things which humans or animals do well but which conventional computers often do poorly. Deep neural networks are neural networks with multiple layers of neurons, which allows them to learn more complex patterns in data than traditional neural networks. They are used in a variety of applications, including image recognition, natural language processing, and machine translation. Neural networks exhibit characteristics and capabilities not provided by any other technology.

The human brain is a complex computing system, capable of thinking, remembering, storing complex patterns, and solving problems. The brain contains approximately 100 billion neurons (fundamental cellular units of the brain's nervous system) that are densely interconnected with hundreds – perhaps thousands – of connections per neuron.

A neuron is a simple processing unit, receiving and combining signals from other neurons through input paths called dendrites. A schematic of neuronal anatomy is shown in Figure 4.1. Input signals come through dendrites and synaptic clefs (functioning like gates) where the information is transmitted chemically across a synapse and electrically inside the soma (body) of the neuron.

The combination of all input signals to a neuron is called *activation*. If the combined signals from all the dendrites are strong enough, the neuron "fires," producing an output signal along a path called the *axon*. The axon splits, connecting to hundreds or thousands of dendrites (input paths) of other neurons through synapses (junctions containing a neurotransmitter fluid controlling the flow of signals) located in the dendrites.

Transmission of the signals across the synapses is electro-chemical in nature, and the magnitudes of the signals depend upon the synaptic strengths of the synapses. The strength or conductance (the inverse of resistance) of a synaptic junction is modified as the brain

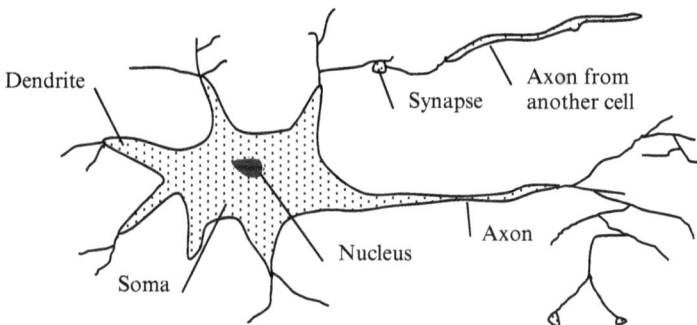

Figure 4.1. Sketch of a neuron showing its components.

"learns." The synapses are the basic "memory units" of the brain, and adjustment of these synapses constitutes learning.

It should be noted that Figure 4.1 is a schematic diagram of only a certain type of neuron found in the cerebral cortex. Many different types of neurons exist in the central nervous system. Some have very different structures. They specialize in different functions. For instance, there are neurons sensing temperature or pressure (force), taste, hearing, transmitting visual data to the brain or activating muscular complexes to achieve movement, grasping, attention direction or the recall of memories. But for computational purposes, the canonical structure we will mostly working with is that of Figure 4.1.

Computer simulation of brain functions usually takes the form of a network of artificial neurons, often called neurons, but sometimes called nodes, processing elements or neurodes. Figure 4.2 shows a simple diagram of an artificial neuron that is analogous to the neuron shown in Figure 4.1, in that it has many inputs (dendrites), combines (sum up) the values of the inputs, and adjusted their weights (synaptic strengths) in response to stimuli. The sum is called the *activation* impinging on the neuron, that is,

$$I = \sum_{i=1}^{n} x_i w_i \tag{4.1}$$

This sum is then subjected to a nonlinear filter, often called an *activation function*, typically a sigmoidal-type function like,

$$\Phi(I) = \frac{1}{1 + e^{-\alpha I}} \tag{4.2}$$

where, α is the *learning rate*, because it adjusts the speed with which weights can be adjusted to learn certain patterns, I is the activation of the neuron and $\Phi(I)$ is the activation function that controls the output in accordance with the prescribed nonlinear relationship.

If the activation function is a threshold function, output signals are generated only if the activation, that is the sum of the weighted inputs, exceeds the threshold value. If the activation function is a continuous nonlinear (or linear) relationship, the output is a continuous function of the combined input.

The most commonly used activation function is the sigmoid function which changes smoothly from zero (for large negative values)

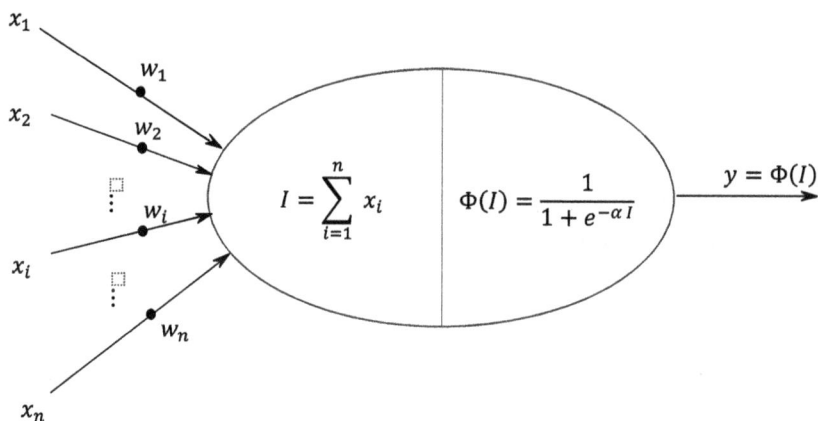

Figure 4.2. Schematic representation of an artificial neuron.

to one (for large positive values) with a value of $1/2$ for zero input. The output axon of an artificial neuron branches out and becomes the input to many other processing elements. These signals pass through connection weights (synaptic junctions) that correspond to the synaptic strength of the neural connections. The input signals to the artificial neuron are modified by the connection weights prior to being summed as shown in Figure 4.2.

4.3 Learning

ML in its broadest sense, uses a number of technologies that includes, but is not limited to, neural networks, genetic algorithms, fuzzy logic, cellular automata, and support vector machines. Interestingly, many of these technologies have their origins in biological or behavioral phenomena related to humans or animals, and analogues of these technologies exist in many human and animal systems. Hence the confidence that AI can achieve tasks that we do not necessarily know how animals do them, yet the fact they are done leads to believe that if we have enough data of their behavior, AI can create models of what's done and maybe discover causal associations that go beyond simple correlation of variables.

ML often involves two or more of these individual technologies used either in series or integrated in a way to produce advantageous results through synergistic interactions. For instance, we can see this in neural and fuzzy synergisms working together, called neurofuzzy systems, a synergism that has been developed for application to various industrial facilities including nuclear power plants (Uhrig and Tsoukalas, 1999).

In data and/or information processing, the objective is generally to gain an understanding of the phenomena involved and to evaluate relevant parameters quantitatively. Patterns associated with specific modes of behavior or characteristics can then be identified. This is usually accomplished through "modeling" of the systems, either experimentally (using data) or analytically (using mathematics and physical principles). Many hybrid systems are used to relate experimental data to system models. Once we have a model of a system, we can carry out various procedures (e.g., sensitivity analysis, statistical regression, etc.) to give insight into the nature of the system behavior. These analyses, in turn, can be used to enhance mathematical and physical models. Data are used to train neural networks and fuzzy logic systems to model the input-output relationships of the systems involved, while genetic algorithms are used to identify near optimal conditions (e.g., best choice of inputs, peak performance based on several criteria, etc.) on a global scale.

As discussed in the previous section, artificial neural networks are defined as computer processing system consisting of many neurons joined together in a structure inspired by the cerebral cortex of the brain. These processing elements are usually organized in a sequence of layers, with full connections between layers as shown in Figure 4.3. Typically, there are three (or more) layers: an input layer where data are presented to the network through an input buffer, an output layer with a buffer that holds the output response to a given input, and one or more intermediate or "hidden" layers. The operation of an artificial neural network involves two processes: *learning* and *recall*.

Learning is the process of adapting the connection weights in response to external stimuli presented at the input buffer. The network "learns" in accordance with a learning rule governing the adjustment of connection weights in response to learning examples applied at the input and output buffers.

Recall is the process of accepting an input and producing a response determined by the geometry and synaptic weights of the network.

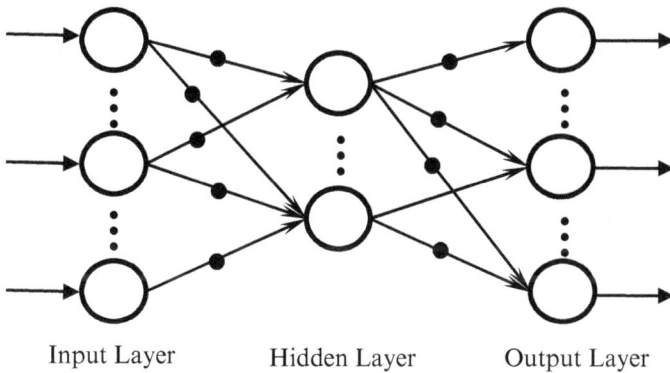

Figure 4.3. Simplified feed forward artificial neural network.

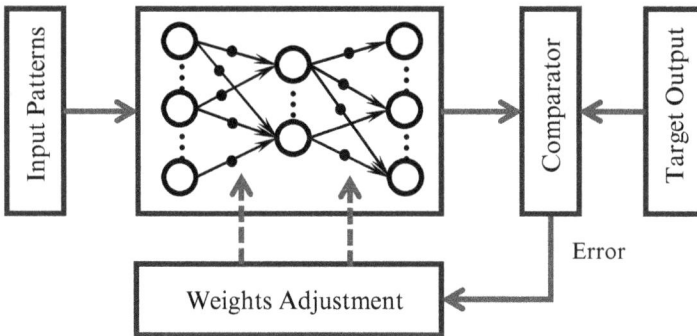

Figure 4.4. An artificial neural network with supervised learning.

Several different kinds of learning commonly are used with neural networks. Perhaps the most common is the so-called *supervised learning* shown schematically in Figure 4.4. Before the learning process begins, all the weights are set to small random values. Then a training input is applied to the input layer, it propagates through the network and produces an output. This output is compared with the desired output to produce an error signal, which, in turn, is the input to the weight adjusting process. Various learning algorithms (discussed below) are used to adjust the weights incrementally. When the input is applied again to the input buffer, it produces an incrementally different output, which again is compared with the desired output and again produces a second error signal. This iterative process continues until the output of the artificial neural network is substantially equal to the desired output, and the error approaches zero or an irreducible minimum. At this point, the network is said to have been "trained."

Through the various learning algorithms, the network gradually configures itself to achieve the desired input-output relationship or "mapping." Supervised learning is often implemented with a technique called backpropagation, the most common method of learning; however, unsupervised learning, as implemented in Kohonen networks, is also used in many situations.

The most common learning algorithms are as follows:

- **Hebbian learning:** Hebbian learning occurs when a connection weight on an input path to a PE is incremented if both the input is high (large) and the desired output is high. This is analogous to the biological process in which a neural pathway is strengthened each time it is used.
- **Delta-rule learning:** Delta-rule learning (sometimes called mean square error learning) occurs when the error signal (difference between the desired output and the actual output) is minimized using a least-squares process. Backpropagation is the most common implementation of Delta-rule learning and probably is used in at least 75% of neural network applications.
- **Competitive learning:** Competitive learning occurs when the processing elements compete; only the processing element yielding the strongest response to a given input can modify itself, becoming more like the input. In all cases, the final values of the weighting functions constitute the "memory" of the neural network.

During the learning phase, the discrepancy between an actual output of a neuron q in the output layer of the network and the known output or target T for that particular input, is called the *total error* ε. The total error is a function of all the weights of the neural network which are modified in the learning phase. In other words, when the total error is zero, all weights have the right values, they cease to be modified and therefore there is no need for further learning. When the error is not zero, the modifications of the proceed in accordance with the contribution of each individual weight to the total error. This is measured by the partial derivative of the total error (squared to simplify the process) with respect to a particular individual weight. This is called the Delta rule of learning, referring to small differential adjustments in the weights of the Adaline machine invented by Bernard Widrow and Ted Huff. Adaline was probably the earliest machine with learning capabilities. Because we will modify a weight with an increment (or decrement based on the sign) we use the square of the total error, that is,

$$\varepsilon^2 = [T - y_q]^2 \qquad (4.3)$$

where T is a target (known) value, and y_q is the output of a neuron labeled as q in the output layer.

The Delta learning rule indicates that the change in any weight w_{ij} is proportional to the rate of change of the square error with respect to that weight, i.e.,

$$\Delta w_{ij} = -\eta \frac{\partial \varepsilon^2}{\partial w_{ij}} \qquad (4.4)$$

where, η is a constant of proportionality called "learning rate." To evaluate this partial derivative, we use the chain rule of differentiation in a cascade of differentiations going from the output layer of neurons to the middle layers and eventually the weights connecting the input layer with the middle (also called "hidden") layers. The learning process is iterative where if t is an iteration step, the weights in $t+1$ would be updated according to the Delta learning rule, that is,

$$w(t + 1) = w(t) + \Delta w_{ij} = w(t) - \eta \frac{\partial \varepsilon^2}{\partial w_{ij}} \qquad (4.5)$$

This is a universal approach to supervised learning which from the attribution of total error to weights in a direction opposite of the forward movement of computation, is called "back propagation of error" (Tsoukalas and Uhrig, 1997).

In the recall process, a neural network accepts a signal presented at the input layer, and then produces at the output layer a response that has been determined by the "training" of the network. The simplest form of recall occurs when there are no feedback connections between layers or within a layer (i.e., the signals flow from the input buffer to the output buffer in a process called "feed forward" information flow). In this type of network, the response is produced in one iteration cycle. When neural networks do have feedback connections, the signal reverberates around the network, across or within layers, until some convergence criteria have been met and a steady-state signal can be presented to the output buffers.

The characteristics that make neural network systems different from traditional computing and AI are: (1) learning by example, (2) distributed associative memory, (3) fault tolerance, and (4) pattern recognition. Let us look at these characteristics.

4.3.1 *Distributive associative memory*

The memory of an neural network is both distributive and associative. "Distributive" means that the storage of a unit of knowledge is distributed across many memory units (connection weights) in the network and shares these memory units with all other items of knowledge stored in the network. "Associative" means that when the trained network is presented with a partial input, the network will choose the closest match, in a least squares sense, to that input in its memory and will generate an output that corresponds to the full output.

4.3.2 *Fault tolerance*

Traditional computer systems are rendered useless by any memory damage; however, neural computing systems are fault tolerant in that if some neurons are destroyed or disabled or have their connections altered incorrectly, the behavior of the network is changed only slightly. As more neurons are destroyed performance degrades gradually; i.e., the network performance suffers, but the system does not fail catastrophically. This behavior is possible because the information is not contained in any single memory unit, but is distributed among many connection weights of the network. Such arrangements are well-suited for energy systems where failure may introduce difficult problems or be unacceptable (e.g., in nuclear power plants, power grid, pipelines, and space probes).

4.3.3 *Pattern recognition*

Pattern recognition is the ability to match large amounts of input information simultaneously and generate a categorical or generalized output. It requires that the network provide a reasonable response to noisy or incomplete inputs. Experience shows that neural networks are very good pattern recognizers which also have the ability to learn and build unique structures for a particular problem.

4.4 Recurrent Neural Networks

Of particular interest are recurrent neural networks due to their role in LLMs as well as in forecasting problems. The backpropagation neural networks previously discussed are strictly "feed-forward" networks in which there are no feedbacks from the output of one layer to the inputs of the same layer or earlier layers of neurons. However, such networks have limited capacity to encode a longer-term memory since the output at any instant is dependent entirely on the inputs and the weights at that instant.

What if we wanted to capture a pattern that depends on inputs from several instances of the past? This is what we are faced with in forecasting problems where a window of past measurements has most of the information needed to predict future values. Typically, the window is not infinite but it is of a size appropriate with the dynamics of the process. An information criterion is used in such cases to estimate the size of the window. For instance, for data that has economic influence, the Akaike information criterion is used, in effect producing the length of the past that contains most of what is needed to predict the future (Tsoukalas and Uhrig, 1997).

An analogous problem exists in the processing of natural language. Suppose that we have 20,000 words indexed in some fashion and we want to make an attribution of meaning to a particular word. For instance, the word "table" may refer to an object in a restaurant, but also to a format of presenting data in a book. Which one is the proper meaning may be inferred if we examine the word "table" in connection with a window of 10 different words that surround it in text. This is the core idea behind LLMs where GPU and TPUs can process structures with many words in a manner that can result in rich semantic attributions that create the impression of a dialogue where the AI interlocutor adapts to the human user prompting AI with questions that allow the bot to not only delineate context but also learn about the educational level and interests on the human side (Tsoukalas, 2023).

There are numerous situations (e.g., when dynamic behavior or semantics are involved) where it is advantageous to using feedback in neural networks. When the output of a neuron is fed back into a neuron in an earlier layer, the output of that neuron is a function of both the inputs from the previous layer at time t and its own output that existed at an earlier time, i.e., at time $(t - \Delta t)$ where Δt is the time for one cycle of calculation. Hence, such networks exhibit characteristics similar to short-term memory, because the output of the network depends on both current and prior inputs.

Although virtually all neural networks that contain feedback could be considered as recurrent networks, we illustrate their workings with backpropagation for learning through an example from (Tsoukalas and Uhrig, 1997).

Let us consider the elementary feedforward network shown in Figure 4.5(a) where the input, middle and output layers each have only one neuron, and neuron h is an input neuron that instantaneously sends the input x to the neuron p in the hidden layer. When the input $x(0)$ is applied to the input, the outputs of neurons p and q at time $(t = 0)$, that is $v(0)$ and $y(0)$ respectively, are

$$v(0) = \Phi[w_{ij}x(0)] \tag{4.6}$$

$$y(0) = \Phi[w_{jk}v(0)]$$
$$= \Phi\{w_{jk}\{\Phi[w_{ij}x(0)]\}\} \tag{4.7}$$

where Φ is the activation function (usually a sigmoidal function) and (0) indicates the value at the start the iteration process, that is $t = 0$.

For non-recurrent networks (no feedback) this relationship remains valid for times $(t = 1, 2, 3, \ldots, n, \ldots, N)$ etc. Hence, Eqs. (4.6) and (4.7) become

$$v(n) = \Phi[w_{ij}x(n)] \tag{4.8}$$

$$y(n) = \Phi[w_{jk}v(n)]$$
$$= \Phi\{w_{jk}\{\Phi[w_{ij}x(n)]\}\} \tag{4.9}$$

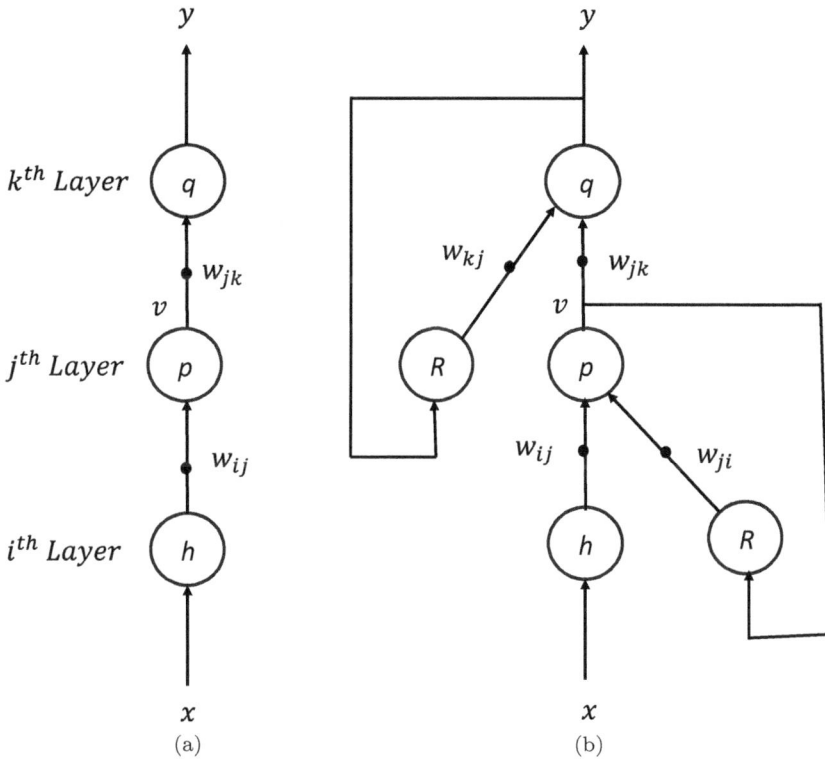

Figure 4.5. Simple neural networks without and with recurrent feedback. (a) Non-recurrent network and (b) recurrent neural network.

Source: Tsoukalas and Uhrig, 1997.

To make this network recurrent, we add feedbacks from the output neuron to the middle layer and from the middle neuron to the input layer through the recurrent neurons labeled R and the corresponding weights w_{kj} and w_{ji} respectively as shown in Figure 4.5(b).

Since the outputs to neurons p and q must exist before there can be any feedback, and the process proceeds step by step, the feedback term of neuron p will not come into play until time $t = 1$ and the feedback for neuron q will not come unto play until time $t = 2$. Hence, the Eq. (4.6) for $y(0)$ is valid for a recurrent network for time $t = 0$, but the feedback terms must be added for all subsequent times.

The output of the network for neurons p and q at time $t = 1$ are

$$v(1) = \Phi\{w_{ij}[x(1)] + w_{ji}[v(0)]\}$$
$$= \Phi\{w_{ij}x(1) + w_{ji}\{\Phi[w_{ij}x(0)]\}\} \tag{4.10}$$
$$y(1) = \Phi\{w_{jk}v(1) + w_{kj}y(0)\}$$
$$= \Phi\{w_{jk}\{\Phi\{w_{ij}\{x(1) + w_{ji}\{\Phi[w_{ij}x(0)]\}\}\}$$
$$+ w_{kj}\{\Phi\{w_{jk}\Phi[w_{ij}x(0)]\}\}\}\} \tag{4.11}$$

For time $t = 2$, the outputs of neurons p and q are

$$v(2) = \Phi\{[w_{ij}x(2)] + [w_{ji}v(1)]\}$$
$$= \Phi\{[w_{ij}x(2)] + w_{ji}\{\Phi[w_{ij}x(1)] + w_{ji}\{\Phi[w_{ij}x(0)]\}\}\} \tag{4.12}$$
$$y(2) = \Phi\{[w_{jk}v(2)] + [w_{kj}y(1)]\}$$
$$= \Phi\{w_{jk}\{\Phi[w_{ij}x(2)] + w_{ji}\{\Phi[w_{ij}x(1)] + w_{ji}\{\Phi[w_{ij}x(0)]\}\}\}$$
$$+ w_{kj}\{\Phi\{w_{jk}\{\Phi[w_{ij}x(1)] + w_{ji}\{\Phi[w_{ij}x(0)]\}\}\}$$
$$+ w_{kj}\{\Phi\{w_{jk}\{\Phi[w_{ij}x(0)]\}\}\}\}\} \tag{4.13}$$

In the third learning iteration, that is time $t = 3$, the outputs of neurons p and q are

$$v(3) = \Phi\{[w_{ij}x(3)] + [w_{ji}v(2)]\}$$
$$= \Phi\{[w_{ij}x(3)] + w_{ji}\{\Phi\{[w_{ij}x(2)] + w_{ji}\{\Phi[w_{ij}x(1)]$$
$$+ w_{ji}\{\Phi[w_{ij}x(0)]\}\}\}\}\} \tag{4.14}$$
$$y(3) = \Phi\{[w_{jk}v(3)] + [w_{kj}y(2)]\}$$
$$= \Phi\{w_{jk}\{\Phi\{[w_{ij}x(3)] + w_{ji}\{\Phi\{[w_{ij}x(2)] + w_{ji}\{\Phi[w_{ij}x(1)]$$
$$+ w_{ji}\{\Phi[w_{ij}x(0)]\}\}\}\}\}\} + w_{kj}\{\Phi\{w_{jk}\{\Phi[w_{ij}x(2)]$$
$$+ w_{ji}\{\Phi[w_{ij}x(1)] + w_{ji}\{\Phi[w_{ij}x(0)]\}\}\}\}$$
$$+ w_{kj}\{\Phi\{w_{jk}\{\Phi[w_{ij}x(1)] + w_{ji}\{\Phi[w_{ij}x(0)]\}\}\}$$
$$+ w_{kj}\{\Phi\{w_{jk}\{\Phi[w_{ij}x(0)]\}\}\}\}\}\} \tag{4.15}$$

Note that the Eq. (4.15) for $y(3)$ has $x(0)$, $x(1)$, $x(2)$ and $x(3)$ as inputs. An equation for $y(4)$ would add $x(4)$ to the list of inputs; $y(5)$ would add $x(5)$; etc. Furthermore, an examination of the terms of the equation for $y(3)$ indicates that the magnitudes of the earlier inputs decrease when later inputs are added. This is seen more readily if we assume that the activation function is linear, i.e.,

$$\Phi\{x\} = x \tag{4.16}$$

Then the equations for $y(0)$, $y(1)$, $y(2)$ and $y(3)$ become

$$y(0) = w_{jk}w_{ij}[x(0)] \tag{4.17}$$

$$y(1) = w_{jk}w_{ij}\{[x(1)] + [w_{ji} + w_{kj}][x(0)] \tag{4.18}$$

$$y(2) = w_{jk}w_{ij}\{[x(2)] + [w_{ji} + w_{kj}][x(1)]$$
$$+ [w_{ji}^2 + w_{ji}w_{kj} + w_{kj}^2][x(0)] \tag{4.19}$$

$$y(3) = w_{jk}w_{ij}\{[x(3)] + [w_{ji} + w_{kj}][x(2)]$$
$$+ [w_{ji}^2 + w_{ji}w_{kj} + w_{kj}^2][x(1)]$$
$$+ [w_{ji}^3 + w_{ji}^2 w_{kj} + w_{ji}w_{kj}^2 + w_{kj}^3][x(0)]\} \tag{4.20}$$

Since the weights are usually less than one, the increasing power of the weights for the earlier terms [w^2 for $x(3)$, w^3 for $x(2)$, w^4 for $x(1)$, and w^5 for $x(0)$] causes the coefficients to decrease rapidly. When the sigmoidal (or any nonlinear activation function) is used, the presence of multiple nonlinear activation functions in the above equations *for* $y(0)$, $y(1)$, $y(2)$, and $y(3)$, each with a maximum value of 1 reduces the early terms much faster than for a linear activation function. As later inputs are introduced, the influence of the earlier terms become negligible. This reduced weighting of the earlier terms is analogous to the decrease of influence for earlier values in a convolution transformation.

The complexity introduced by feedback connections, even for the elementary system of Figure 4.5(b), is readily apparent. For networks more complicated than the elementary system in Figure 4.5(b), the same principles are applied, but the complexity grows even more rapidly. This is where GPU and TPUs come to the rescue. In LLMs, a window where a vector of properly labeled 10 words surrounding, for instance, the word "table," allows the network to discriminate between a restaurant table and a book table. Although the increase in complexity is often compensated for because the feedback drastically reduces the number of cycles needed to train a neural network when the window is increased to the words found in a paragraph or a page, the high-performance hardware is of great importance. Feedback can often be used advantageously to speed up the training of a neural network and to avoid local minima. Indeed, it is sometimes possible to train a neural network after feedback has been added whereas it may not have been previously possible to train it to the desired low level of error. However, capturing dynamic behavior in a model is the most common justification for the use of feedback, and recurrent neural networks are almost always used for dynamic signals.

Long Short-Term memory (LSTM) networks are based on recurrent neural networks and are commonly used in time series forecasting tasks, due to their efficiency in processing sequential data with long-term dependencies.

LSTM networks address issues with very small or large error gradients often encountered in traditional recurrent networks. LSTM cells consist of *input, forget,* and *output gates*, which control the flow of information. The input gate determines the amount of new information that should be stored in the cell, while the forget gate decides what information should be discarded. The output gate regulates the flow of information towards the next layer or output (Pantopoulou *et al.*, 2025). The LSTM cell calculates a hidden state H_t and a current state C_t as follows,

$$H_t = \sigma(x_t U_o + H_{t-1} W_o) \cdot \tanh(C_t) \qquad (4.21)$$

$$C_t = \sigma(x_t U_f + H_{t-1} W_f) C_{t-1}$$
$$+ \sigma(x_t U_i + H_{t-1} W_i) \tanh(x_t U_c + H_{t-1} W_c) \qquad (4.22)$$

Here x_t is the input to the cell, H_{t-1} and C_{t-1} are hidden and current states from the $t-1$ time step, σ is a sigmoid activation function, U and W are weight matrices. Subscripts f, i, o, and c refer to forget, input, output gates and current state, respectively.

Recent studies have investigated performance of LSTM networks in various applications, including nuclear reactor control, fiber-optic sensor monitoring, failure detection and remaining useful lifetime estimation, detection of wind turbine blade icing, short-term forecasting of solar energy systems, and water distillation.

LSTM-based strategies have moreover been used for the detection of thermocouple sensor failures from time series analysis. In a recent work, the authors investigated the performance of LSTM and transfer learning LSTM (TL-LSTM) in monitoring and estimating thermocouple time series in water and Galinstan fluids at room temperature and ambient pressure (Pantopoulou *et al.*, 2022b; Prantikos *et al.*, 2023).

LSTMs for predicting time series of a thermocouple in each loop were constructed by training on prior history of the same sensor, while TL-LSTMs were constructed by training on the history of a correlated sensor. The benchmarking of LSTM performance indicated that monitoring can be performed in real time. Another study benchmarked the performance of LSTM and statistical Auto Regressive Integrated Moving Average (ARIMA) models in temporal forecasting of distributed temperature sensing (Pantopoulou *et al.*, 2025). The benchmarking study investigated temperature forecasting in Vanilla, zero shot forecasting and TL use cases. In Vanilla use case, predictions of fiber optic sensors (FOS) temperature were made with LSTM and ARIMA trained on the data from the same FOS that was used for testing. In ZSF use case, forecasting of one FOS was made with ARIMA and LSTM trained on another FOS. In TL use case, LSTM developed in ZSF was re-trained for improved performance. The study showed that ARIMA outperforms LSTM in all use cases; however, development of ARIMA requires several data pre-processing with statistical tests to determine model parameters. While ARIMA proved to be relatively insensitive to correlations between training and testing datasets, LSTM performance depends both on the correlation between the training and testing data and on the volume of the training data.

LSTM networks have moreover been studied in conjunction with TL to validate temperature time series in high temperature molten salt loops (Pantopoulou *et al.*, 2022a), and in conjunction with autoencoder neural networks to monitor flow rate measurements (Pantopoulou *et al.*, 2024).

4.5 Language and Computation

The advent of large language models, or LLMs, is based on recurrent neural networks capturing nuances of meaning in a word based on a neighborhood of surrounding words in a document. With ever more capable hardware (GPU and TPUs) the neighborhood has grown significantly without being overwhelmed by the complexities of feedback in the networks. Yet, another feature of language inherent in the symbolization of the world merits attention. This is fuzziness.

Fuzziness is found in the construction of groups or sets when we place within their linguistic perimeter elements with varying degrees of memberships and hence with an excess of meanings (Tsoukalas, 2023). For instance, a rather small number of words for color, e.g., *green*, *blue* and *red* suffice to describe the millions of colors we can see and paint with. There is nothing fuzzy about the colors themselves. They are what they are. Fuzziness is a feature of the *symbolization of the World*. It has to do with how a mind (or AI program) uses symbols to enable us to collaborate and coordinate narratives and activities successfully.

In mapping the physical world into a milieu of narratives the notion of a set is (often implicitly) used to define values and variables, their mappings, ranges and interactions. Whereas in classical set theory one finds a rather strict sense of membership, that is, an element either belongs or does not belong to a set, an all-or nothing situation, in *fuzzy sets* infinite degrees of membership are allowed.

This is mathematically indicated through *a membership function,* i.e., a function unique for each and every fuzzy set A, which maps every element of the universe x of discourse X to the interval $[0, 1]$, formally written as $\mu_A(x) : X \to [0, 1]$. Membership functions may represent an individual's (subjective) notion of a vague class, e.g., *acceptable performance, small contribution to system stability, major improvement, big benefit*, etc. In designing and operating controllers or automatic decision-making tools, for example, modeling such notions is a very important task. Membership functions may also be determined on the basis of statistical data or through the aid of neural networks; the synergistic relation between neural networks and fuzzy logic towards this end is described in (Tsoukalas and Uhrig, 1997; Tsoukalas, 2023). It should be emphasized, however, that although membership functions may be subjective, they are not assigned arbitrarily, but rather on the basis of application-specific criteria.

If X is a universe of discourse and x a particular element of X, then a fuzzy set A defined on X may be written as the *union* of all

$\mu_A(x)/x$ singletons, i.e.,

$$A = \int_x \mu_A(x)/x \qquad (4.23)$$

where the *integral* sign in Eq. (4.18) indicates the *union* of all $\mu_A(x)/x$ singletons.[2] Consider, for example, the fuzzy set *small numbers* defined (subjectively) over the set of non-negative real numbers through a continuous membership function $\mu_A(x)$. The set A may be written as

$$A = \int_{x \geq 0} \mu_B(x)/x = \int_{x \geq 0} \left[\frac{1}{1 + \left(\frac{x}{5}\right)^3} \right] \Big/ x \qquad (4.24)$$

The *union* of two fuzzy sets A and B defined over the same universe of discourse X, is a new fuzzy set $A \cup B$ also on X, with membership function which is the maximum of the grades of membership of every x to A and B, i.e., $\mu_{A \cup B}(x) \equiv \mu_A(x) \vee \mu_B(x)$.

The *union* of two fuzzy sets is related to the logical operation of *disjunction* (*OR*) in logic. The *intersection* of two fuzzy sets A and B is a new fuzzy set $A \cap B$ *with* membership function which is the minimum of the grades of every x in X to the sets A and B, i.e., $\mu_{A \cap B}(x) \equiv \mu_A(x) \wedge \mu_B(x)$.

The *intersection* of two fuzzy sets is related to *conjunction* (*AND*) in logic.

The *complement* of a fuzzy set A is a new fuzzy set, A, with membership function $\mu_{\bar{A}}(x) \equiv 1 - \mu_A(x)$.

Fuzzy set *complementation* is equivalent to *negation* (*NOT*) in logic. A particular type of fuzzy set is a *fuzzy number*, defined as a fuzzy set on the set of real numbers (or its subsets, naturals, integers, etc.) which is *normal* (i.e., it has membership equal to one for at least one element) and *convex*.

In fuzzy systems, *relations* possess the computational potency and significance that *functions* have in conventional approaches.

---

[2]Note that the integral sign is not the same as the integral sign of differential and integral calculus. It is used here in the sense that the integral sign is used in set theory which is to indicate the *sum* or *union* of individual *singletons*.

Fuzzy *if/then* rules and their aggregations, known as *fuzzy algorithms* or *fuzzy systems*, both of central importance in engineering applications, are fuzzy relations in linguistic disguise. Fuzzy relations may be thought of as fuzzy sets defined over high-dimensional universes of discourse. Crisp relations are defined over the *Cartesian product* or *product space* of two or more sets.

The *Cartesian product* $X \times Y$ of two sets X and Y is the set of all ordered pairs (x, y) with x in X and y in Y. Relations may also be thought of as *mappings*. *Functions* are *mappings* as well. In fuzzy systems, however, we deal with relations of a more general nature. A function performs what is called a *many-to-one mapping*, that is, many elements are associated with one (and only one) element but not vice versa.

For example, if the mapping is done between x's and y's in the $X \times Y$ plane, we may have more than one x mapped to the same y but not the other way around. In fuzzy relations, however, many-to-many mappings are performed. Many x's can be associated with a single y and vice versa. Many y's can also be associated with a single x. The importance of this abstract-sounding distinction in terms of engineering and computational applications cannot possibly be overstated. A fuzzy relation R can be written as

$$R = \int_{X \times Y} \mu_R(x, y)/(x, y) \qquad (4.25)$$

The notion of composition of fuzzy relations is very important in fuzzy systems. A fuzzy system is the transcription of a narrative as a collection of fuzzy *if/then* rules, each rule being a fuzzy relation and the problem of inferencing (or evaluating them with specific inputs) is mathematically equivalent to composition.

There are several types of composition. By far the most common in engineering applications is *max-min composition*, but also *max-star*, *max-product*, and *max-average* are used. Given two fuzzy relations R_1 and R_2 the composition of the two is a new relation given by

$$R_1 \circ R_2 \equiv \int_{X \times Y} \bigvee_y [\mu_{R_1}(x, y) \wedge \mu_{R_2}(y, z)]/(x, z) \qquad (4.26)$$

where, the symbol "∘" stands for *max-min composition* of relations R_1 and R_2, $\vee$ stands for maximum and $\wedge$ stands for minimum.

Example 4.1. Max-Min Composition of Fuzzy Relations.
We are given two fuzzy relations R_1 and R_2

$$R_1 = \begin{bmatrix} 1.0 & 0.3 & 0.9 & 0.0 \\ 0.3 & 1.0 & 0.8 & 1.0 \\ 0.9 & 0.8 & 1.0 & 0.8 \\ 0.0 & 1.0 & 0.8 & 1.0 \end{bmatrix}, \quad R_2 = \begin{bmatrix} 1.0 & 1.0 & 0.9 \\ 1.0 & 0.0 & 0.5 \\ 0.3 & 0.1 & 0.0 \\ 0.2 & 0.3 & 0.1 \end{bmatrix}$$

What is the max-min composition?

Solution: Using Eq. (4.26), the max-min compositions is a new relation. The operation is actually similar to matrix multiplication with ($\vee$) being treated like *summation* (+) and ($\wedge$) being treated like *multiplication* ($\cdot$), i.e.,

$$R_1 \circ R_2 = \begin{bmatrix} 1.0 & 0.3 & 0.9 & 0.0 \\ 0.3 & 1.0 & 0.8 & 1.0 \\ 0.9 & 0.8 & 1.0 & 0.8 \\ 0.0 & 1.0 & 0.8 & 1.0 \end{bmatrix} \circ \begin{bmatrix} 1.0 & 1.0 & 0.9 \\ 1.0 & 0.0 & 0.5 \\ 0.3 & 0.1 & 0.0 \\ 0.2 & 0.3 & 0.1 \end{bmatrix} \qquad (4.1.1)$$

To evaluate Eq. (4.1.3) we proceed, like in matrix multiplication, by forming the pairs of minima of each element in the first row of membership matrix R_1 with every element in the first column of membership matrix R_2.

For example, to obtain the first element, that is, first row and first column, we perform the following operations

$$\begin{bmatrix} 1.0 & 0.3 & 0.9 & 0.9 \end{bmatrix} \circ \begin{bmatrix} 1.0 \\ 1.0 \\ 0.3 \\ 0.2 \end{bmatrix}$$

$$= [1.0 \wedge 1.0] \vee [0.3 \wedge 1.0] \vee [0.9 \wedge 0.3] \vee [0.0 \wedge 0.2]$$

$$= 1.0 \vee 0.3 \vee 0.0$$

$$= 1.0 \qquad (4.1.2)$$

We repeat this procedure for all rows and columns and the result is the membership matrix of the composed relation $R_1 \circ R_2$ given by

$$R_1 \circ R_2 = \begin{bmatrix} 1.0 & 1.0 & 0.9 \\ 1.0 & 0.3 & 0.5 \\ 0.9 & 0.9 & 0.9 \\ 1.0 & 0.3 & 0.5 \end{bmatrix} \qquad (4.1.3)$$

Fuzzy formulations offer an alternative and often complementary language to conventional (analytic) approaches to modeling systems (involving differential or difference equations). Informal linguistic narratives used by humans in daily life as well as in the performance of skilled tasks, such as control of industrial facilities, troubleshooting, aircraft landing and so on, are usually the starting point for the development of fuzzy systems. Although fuzzy formulations are based in a human-like language, they have rigorous mathematical foundations involving fuzzy sets and relations. They encode knowledge about a system in statements of the form

> *if* (a set of conditions are satisfied)
>
> *then* (a set of consequences can be inferred)

For instance, in control the desirable behavior of a system may be formulated as a collection of rules combined by the connective *ELSE* such as

> *if error* is *ZERO AND* $\Delta error$ is *zero then* Δu is *ZERO   ELSE*
>
> *if error* is *PS AND* $\Delta error$ is *zero then* Δu is *NS   ELSE*
>
> . . .
>
> *if error* is *SMALL AND* $\Delta error$ is *ns then* Δu is *BIG*   (4.27)

where, error, $\Delta error$ (change in error) are linguistic variables describing the input to a controller and, Δu a linguistic variable describing the change in output. A *linguistic variable* is a variable whose arguments are fuzzy numbers (and more generally words modeled by fuzzy sets) which we refer to as *fuzzy values*. For example, in the rules above the *fuzzy values* of the linguistic variable *error* are *ZERO*, *PS* (*positive small*) and *SMALL*, the values of $\Delta error$ are *ZERO* and *NS* (*negative small*) and the values of Δu are *ZERO*, *NS* and *BIG*.

A specific evaluation of a fuzzy variable, for example, "*error* is *zero*" is called *fuzzy proposition*. Individual fuzzy propositions on either *left* (*LHS*) or *right-hand-side* (*RHS*) of a rule may be connected by connectives such as *AND* and *OR* for example, "*error* is *ps AND* $\Delta error$ is *zero*." Individual *if/then* rules are connected with the connective *ELSE* to form a *fuzzy algorithm*. Propositions and *if/then* rules in classical logic are supposed to be either true or false. In fuzzy logic they can be true or false to a degree. In the simplest possible case, the *i*th fuzzy *if/then* rule of a fuzzy system has the canonical form

$$if\ x\ is\ A_i\ then\ y\ is\ B_i \tag{4.28}$$

The analytical form of rule (4.23) is a fuzzy relation $R_i(x, y)$ called the *implication relation* of the rule (there are over 50 possible forms for this relation).

Suppose that we have a function $y = f(x)$ and we want to approximate it by a collection of fuzzy *if/then* rules such as

$$if\ x\ is\ A_l\ then\ y\ is\ B_l \quad ELSE$$
$$if\ x\ is\ A_2\ then\ y\ is\ B_2 \quad ELSE$$
$$\vdots$$
$$if\ x\ is\ A_i\ then\ y\ is\ B_i \quad ELSE$$
$$\vdots$$
$$if\ x\ is\ A_n\ then\ y\ is\ B_n \tag{4.29}$$

where, $A_1, A_2, \ldots, A_i, \ldots, A_n$ are fuzzy numbers on the x-axis and, $B_1, B_2, \ldots, B_i, \ldots, B_n$ are fuzzy numbers on the y-axis. The rules of (4.24) are combined by the connective *ELSE* which could be analytically modeled either as intersection or union depending on the implication relation of the individual rules. We recall that a function is a special kind of relation that associates a unique y with each x. A function performs what is called a *many-to-one mapping*, i.e., several values of x may have the same value of y but not *vice versa*. Most real-world applications, however, involve *many-to-many mappings*.

Sometimes conventional narratives, being overly idealized models of complex systems, may suffer from lack of robustness and exhibit undesirable side-effects. The process of using a fuzzy system is called fuzzy inference. There are two important problems in fuzzy inference. First, given a fuzzy number A' as input to a linguistic description we want to obtain a fuzzy number B' as its output and, second given B' we want to obtain A' (the inverse problem). The first problem is addressed with an inferencing procedure called *generalized modus ponens* (*GMP*), and the second with another inferencing procedure called *generalized modus tollens* (*GMT*). Both *GMP* and *GMT* have their origin in the field of logic and approximate reasoning and analytically they involve composition of fuzzy relations.

In general, fuzzy narratives offer convenient tools for controlling the *granularity* of a narrative.[3] in the sense that they facilitate the choice of appropriate precision levels, that is, levels that application-specific considerations call for. In terms of our example, when we use fuzzy numbers and fuzzy *if/then* rules to describe $y = f(x)$ we have at our disposal a mechanism for reducing the number of rules needed and, hence for controlling the *granularity* of this particular description and the overall cost of computation. In addition, the technology for computing with *if/then* rules has already advanced to the point where fuzzy microprocessors, called *fuzzy chips*, are available.

Chips encoding knowledge in the form of narratives can function as "embedded devices" that is, dedicated processors fine-tuned to the specifics of a component and its environment performing domain-specific computations. Such processors are already deployed in several control and robotics applications with remarkable successes. Of course, software is a commonly used medium for the implementation of fuzzy algorithms on a variety of different computers. However, the advent of high-performance hardware the AI inference support may have a profound impact on the design and operation of LLMs. Fuzzy narratives are of growing importance in delineating the domains of LLMs enabling AI to easily adapt to applications of control, pattern recognition, signal analysis, reliability engineering and ML. The basic ideas however are rather similar and rest on the mathematics of fuzzy sets (Tsoukalas, 2023).

---

[3]By *granularity* we roughly mean the coarseness of a narrative, the level of precision necessary to effectively represent a given system.

As an example of LLMs at work in facilitating the Atomic energy transition consider the problem of managing records pertaining to nuclear regulation (Prantikos *et al.*, 2024). Processing documents related to nuclear reactors presents significant challenges due to the volume and complexity involved. Operators, engineers, and regulatory inspectors often need to navigate through volumes of technical documents to under guidelines, and safety procedures. This documentation might not be organized efficiently, making it challenging to search for relevant information. However, AI information retrieval offers innovative solutions to these challenges.

The Retrieval Augmented Generation (RAG) model was developed to quickly and accurately retrieve relevant information from a vast number of documents, as it combines the robustness of neural network-based language models with the precision of information retrieval techniques. RAG allows users to ask questions in natural language and receive contextually relevant answers by integrating the generative capabilities of Generative Pre-trained Transformer (GPT) models with the document retrieval functionalities exemplified by Dense Passage Retrieval (DPR) models. RAG models excel in tasks requiring external knowledge, such as question answering, by accessing a broader range of information than the training data used during pretraining and fine-tuning of LLMs.

The RAG model has been applied to process the technical documentation of a laboratory that supports the development of next-generation technologies and workforce training for commercialization of sodium fast reactors. The operation of the facility is detailed in hundreds of pages of technical reports. Information search and retrieval for researchers and operators using facility operations documentation is developed. The RAG approach leverages the Transformer architecture to enhance information retrieval and generation capabilities (Gao *et al.*, 2024). Transformers have revolutionized Natural Language Processing by introducing a parallelized architecture that significantly improves speed and scalability (OpenAI *et al.*, 2024). Transformers rely on attention mechanisms instead of sequence-based processing, enabling more efficient handling of complex language tasks (Vaswani *et al.*, 2017).

The Transformer architecture consists of two main components: the *encoder* and the *decoder*. Each of these consists of multiple layers with attention mechanisms and fully connected feed-forward networks. A key feature is self-attention, which calculates attention scores from different parts of the input, allowing the model to weigh the significance of each word independently of its position. This enhances the understanding of context and relationships in language. Attention mechanisms allow the model to focus on different words in the input sequence for each word in the output sequence, efficiently modeling the complex relationships in the text. Multiple attention heads learn different data aspects, providing a nuanced understanding of the context.

Information retrieval in LLMs is based on embedding similarity. Embeddings are dense, low-dimensional vectors representing high-dimensional data. In NLP, embeddings are generated for words, sentences, paragraphs, or entire documents, transforming the raw text into vectors. These vectors capture semantic properties, placing similar meanings close together in the embedding space, enabling complex linguistic relationships to be modeled mathematically.

In information retrieval, embedding similarity allows for efficient searches through large datasets by comparing query embeddings with a database of pre-computed embeddings. Similarity metrics like cosine similarity, Euclidean distance, or Manhattan distance quantify these comparisons, with cosine similarity being most popular for capturing semantic orientation.

Embeddings are generated by training on extensive text corpora using models like Word2Vec, GloVe, Bidirectional Encoder Representations (BERT), and OpenAI. For example, BERT creates contextually rich embeddings by processing words in relation to their surrounding context within sentences. For our application, we are utilizing the "text-embedding-3-large" embedding model developed by OpenAI and the cosine similarity metric. This metric measures the cosine of the angle between two vectors rather than the magnitude. Given two vectors, A and B, each representing a sentence or chunk in the vector space, the cosine similarity, $\cos(\theta)$, between these sentences can be calculated as

$$\cos(\theta) = \frac{A \cdot B}{\|A\| \|B\|} \tag{4.30}$$

where $A \cdot B = \sum_{i=1}^{n} A_i B_i$ is the dot product of the vectors A and B, $\|A\| = \sqrt{\sum_{i=1}^{n} A_i^2}$ and $\|B\| = \sqrt{\sum_{i=1}^{n} B_i^2}$ are the magnitudes of the vectors A and B, respectively. This measure ranges from -1 to 1, where 1 means the two sentence vectors are pointing in the same direction, indicating high similarity, and -1 indicates the two sentence vectors are diametrically opposed, indicating high dissimilarity. A cosine similarity of 0 implies orthogonality or no similarity between the vectors.

The RAG architecture integrates a retrieval system with a generator. Initially, the retrieval system fetches relevant documents or passages from a pre-indexed corpus based on the input query. This retrieval uses dense vector representations to map semantically similar texts to nearby points in the embedding space. After retrieving relevant documents, the context and the initial query are passed to a generator based on the Transformer model.

Figure 4.6 illustrates the RAG architecture. The RAG process begins with the retrieval module converting the query into an embedding using the same method as during pre-processing. One performs a similarity search across the corpus to find the best-matching documents, using techniques like cosine similarity for high-dimensional vector spaces. The top documents are combined with the query to

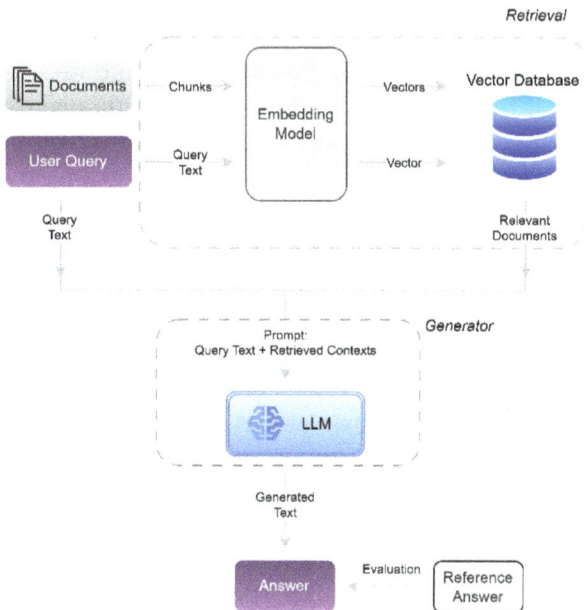

Figure 4.6. LLM architecture for facility documentation.

form a comprehensive input (prompt) for the sequence-to-sequence model. This model treats the combined input as an extended text from which to generate an output, influenced by both the direct input and the learned relationships and patterns from its training.

Basic RAG approach first divides the input documents into chunks, embeds them, and stores them in the index. For a given query, one embeds the query and retrieves the top k-closest embeddings from the index. These embeddings are then used by the LLM to generate the final answer. Small-to-Big RAG combines the advantages of small and large chunks. Small chunks offer high retrieval accuracy because they contain precise information, but they lack sufficient context for the LLM to generate a comprehensive answer. Conversely, large chunks provide more context but can confuse the embedding model, making it difficult to represent the contextual meaning accurately. The Small-to-Big approach uses small chunks for retrieval and larger chunks for generation. The Small-to-Big configuration may be taken in breaking original chunks of size 1024 into 512 and 256, or another breaking them into 512, 256, and 128. The smaller chunk size is the one that is used for retrieval. Overlapping or neighboring small chunks are merged to avoid duplicate information. Sentence-Level approach breaks documents into single-sentence chunks, including a window of sentences before and after each chunk. Similar to Small-to-Big RAG, single sentences enhance retrieval precision, while the surrounding sentences provide additional context for generation. This method improves the generation's performance by expanding the context window around the retrieved sentence.

4.6 Genetic Algorithms

Genetic algorithms as a field of study were developed by John Holland (Holland, 1992) and his students, but its applications to real-world practical problems was almost two decades in developing. In one way or another, the primary purpose of using genetic algorithms is optimization. There is no guarantee that a genetic algorithm will give an optimal solution or arrangement; only that the solution will be near optimal in the light of the specific fitness function used in the evaluation of the many possible solutions generated (Goldberg, 1989).

Genetic algorithms mimic some of the processes of natural evolution. In doing so, some of the inherent features of evolution are utilized in fields far beyond genetics. However, there is no necessity that genetic algorithms as we use them mimic in detail the behavior of the evolutionary process. Indeed, users are free to utilize those features that are useful and discard aspects that seem unimportant in their applications. Since normal evolution processes are quite slow, biased reproduction, based on an aggressive "survival of the fittest" philosophy is used to speed up the evaluation process.

Chromosomes are composed of genes which define the characteristics of the chromosomes and may take on several values called alleles. The position of a gene (its locus) is identified separately from the gene's function. Hence, we can have a particular gene with a locus of position 12 whose allele value is brown. Generally, the strings of artificial genetic systems are analogous to chromosomes in biological systems and are often called chromosomes.

The mechanisms that induce evolution are not well understood, but the features of evolution have been investigated thoroughly. First, evolution takes place on chromosomes, the genetic units that encode the features and structure of living creatures. The specific descriptive features of a living creature is determined by the chromosomes of the previous generation, and evolution influences only these chromosomes, not the living creature from which they came. Since evolution is limited to chromosomes, living creatures do not evolve during their lifetime; their features, which are presumably set at the time of conception, are different from the previous generation only because the chromosomes of their parents changed through evolution.

Natural selection is a process by which nature causes those chromosomes that encode better characteristics (by some criteria) to reproduce more often than those that encode poorer characteristics. Natural selection is the processes that causes genetic algorithms to produce near optimal solutions when the selected chromosome is decoded. This process involves creation of many chromosomes by reproduction, mating (crossover), mutation, and the survival of the chromosomes with the better characteristics. Successive generations of chromosomes improve in quality, provided the criteria used for survival is appropriate. This process is often referred to as Darwinian natural selection or the "survival of the fittest." In nature, this process of evolution occurs over many years, even hundreds or thousands of years. In the computer, the representations of chromosomes can undergo literally thousands of generations of change in a few seconds. Historically, the characteristics of the chromosomes in genetic algorithms have been represented by 0's and 1's. All of Holland's work used this representation, and we will use it here. However, chromosomes can be represented by real numbers, permutations of elements, a list of rules, or other symbols. Binary representation is still the most common representation because the behavior of bit strings is better understood.

Evolution takes place through the process of reproduction which involves mutations and recombination after mixing of the parent's chromosomes to produce a creature that may have entirely different characteristics from the previous generation. The chromosomes of the two parents are mixed by a process called "crossover" in which two new chromosomes are produced, each having some of the characteristics of the two parents. The two parent chromosomes split at some point, and one part of one parent chromosome is exchanged for the corresponding part of the other parent chromosome. The location of the crossover point at which the parents' chromosomes divide is apparently a uniform random process. If one of the new chromosomes obtains 75% of its characteristics from one parent and 25% from the other, the second new chromosome gets 25% form the first parent and 75% from the other.

Both of the resultant chromosomes go into the gene pool (where all the alleles reside) where they either replace poorer quality chromosomes or are discarded. Once a chromosome is discarded, its unique features (good or bad) are lost forever. It is only through the processes of mutation (described below) that such a chromosome might possibly be recreated.

Mutation is a process by which a single component of a chromosome is changed randomly. It occurs in only a very small fraction (typically less than a fraction of a percent) of the chromosomes. It represents an abrupt change in the nature of the chromosome and influences all subsequent generations of chromosomes containing this mutated component. Of course, if this mutation results in a poorer quality chromosome, it will be discarded and lost from the gene pool.

Each population has a gene pool consisting of a large number of chromosomes generated by the process of natural selection.

The choice of which two chromosomes are mated and subject to the crossover process is somewhat random, but the fitter chromosomes are more likely to be selected first. New chromosomes are constantly being reproduced by the mating process described above, and those with better characteristics are retained while those with poorer characteristics are discarded. Generally, but not always, the mixing of chromosomes with quite different characteristics produce better chromosomes. As the process proceeds, the average quality of the gene pool improves because the poorer quality chromosomes are discarded.

The function on which an optimization algorithm operates, i.e., seeking its maximum or minimum, is called the objective function. In neural networks, the objective function to be minimized is the mean square error over the entire training set. In genetic algorithms, the "fitness" is the quantity that determines the quality of a chromosome, from which a determination can be made as to whether it is better or worse than other chromosomes in the gene pool. The fitness is evaluated by a "fitness function" that must be established for each specific problem. This fitness function is chosen so that its maximum value is the desired value of the quantity to be optimized. Its importance cannot be overemphasized, because it is the only real connection between the genetic algorithm and the problem in the real world.

A fitness function must reward the desired behavior; otherwise the genetic algorithm may solve the wrong problem. Fitness functions should be informative and have regularities. However, they need not be low-dimensional, continuous, differentiable, or unimodal.

Before using a genetic algorithm, the solution that is sought has to be coded in a binary string, called a chromosome. A genetic algorithm involves the following steps:

First, a population of randomly generated chromosomes is selected. These chromosomes represent initial guesses of solutions (also called the first generation). The fitness of each individual chromosome is evaluated.

Second, a selection procedure is employed to form the next generation of chromosomes. The selection is a statistical process, that is, a chromosome may or may not be chosen to be present in next generation with its chance proportional to its fitness.

Third, two genetic operators, mutation and crossover, are applied to the population to generate new species. The probability of mutation and crossover is controlled by two parameters: mutation rate and crossover rate.

The benefits of using genetic algorithms are: (i) it is likely to find a near-optimal solution within a small number of iterations and (ii) it does not rely on any *a priori* model of input and output relations, which is better for field operations.

The genetic algorithm is usually used with ill-defined or optimization problems with complex objective functions. Optimal power flow (OPF) is certainly a complex optimization problem with often a multi goal objective function. It can also be classified as an ill-defined problem in terms of classical solutions. General OPF formulation is non-linear, non-convex, mixed continuous/discontinuous optimization problem. Genetic algorithms are capable of broad and local search regardless of the type of objective function.

There is a large number of different genetic algorithms although the basic principles are almost identical. Genetic algorithm is a stochastic optimization algorithm which searches for an optimal solution among possible solutions (genomes) by perturbing the variable space and moving in the direction of decreasing objective function cost. Figure 4.7 shows a generic GA algorithm operation.

For the OPF, it is important to note that by constraining genomes' ranges the OPF constraints are satisfied. It is also important to understand that the genomes can be configured to take continuous values, values increased in steps, or just a value from a set of predefined values. This corresponds to stepwise control of generator outputs, transformer tap settings, etc. A drawback of using GA for optimization is that due to its stochastic nature, each GA OPF run results in a slightly different optimal point. This deficiency is usually corrected with some kind of gradient method correction. However, if a gradient method were available for a particular objective function, we would not need a GA in the first place. Also, the observed differences between the runs usually correspond to a few MWs at most, resulting in sufficient accuracy for the OPF problem.

The basic idea is to use continuous genomes as controlling variables. A power flow is run with the modified set of controls and the objective function is evaluated. The objective function format depends on what we are trying to minimize (or maximize) and can be

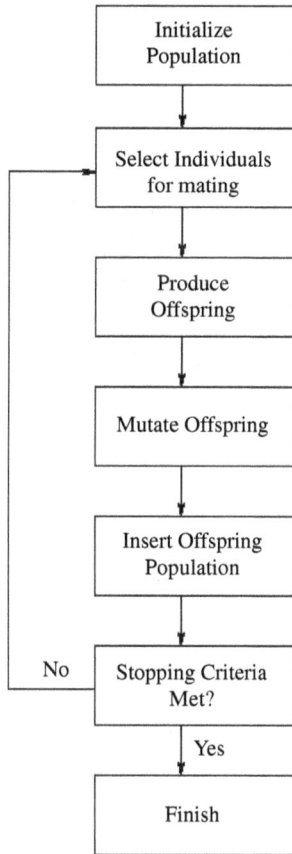

Figure 4.7. Genetic Algorithm.

based on generation cost, losses, congestion, or a combination of some or all of them.

The most obvious and easiest approach would be to replace the Power Flow solver with a more efficient one. This was left to be implemented as an option to be chosen by a user. Another possibility is to parallelize the GA OPF and distribute it over a number of non-dedicated workstations. The third option is to use ML for Unit Commitment and perform the learning phase off-line. Each of the last two approaches has its benefits and drawbacks. Both have been implemented and evaluated (Tsoukalas and Uhrig, 1997).

Problems

Problem 4.1. If a neural network doesn't train to a desired total error goal ($\epsilon \approx 0$) what could be the cause and what should be done?

Problem 4.2. A simple network consists of two neurons in the input layer connected to an output neuron. It is a single layer non-linear network receiving two inputs $x_1 = 2$ and $x_2 = -2$ while the target output should be 0.4. The synaptic weights linking input neurons with output are 0.5 and -0.5 respectively. Derive the update equations. What are the next two outputs, error terms and weight updates if the learning rate is 1.0. How could learning be speeded up?

Note:

$$f(x) = \text{logsig}(x) \quad \text{and} \quad f'(x) = f(x)[1 - f(x)]$$

You may use the table below for convenience.

x	−2.0	−1.50	−1.00	−0.50	0	0.50	1.00	1.50	2.00
$\text{logsig}(x)$	0.1	0.2	0.3	0.4	0.5	0.6	0.7	0.8	0.9

Problem 4.3. Explain how the back propagation of error algorithm works and construct a simple three-layer network (see Figure 4.5(a)) with one neuron per layer to show how it's workings. Design a simple experiment to learn a straight line and discuss what you learned from your experiment.

Problem 4.4. In the recurrent neural network of Figure 4.5(b), the feedback of both loops comes into action at time $t = 1$. If the feedback from the output neuron does not come into action until time $t = 2$, how will Eqs. (4.8) through (4.15) change?

Problem 4.5. Two fuzzy sets A and B are given by

$$A = 0.33/6 + 0.67/7 + 1.00/8 + 0.67/9 + 0.33/10$$
$$B = 0.33/1 + 0.67/2 + 1.00/3 + 0.67/4 + 0.33/5$$

Find the union, intersection and negation of these sets and sketch their membership functions.

Problem 4.6. Given the fuzzy sets A and B in Problem 4.4, construct the implication relation for the rule

$$\text{if } u \text{ is } A \text{ then } v \text{ is } B$$

Using the following impli*cation operators*

Mamdani Min: $\mu_{AB}(u, v) = \mu_A(v) \wedge \mu_B(u)$

Larson Product: $\mu_{AB}(u, v) = \mu_A(v) * \mu_B(u)$

Zadeh max-min: $\mu_{AB}(u, v) = (\mu_A(v) \wedge \mu_B(u))v(1 - \mu_A(v))$

Problem 4.7. A linguistic description is comprised of a single rule

$$\text{if } x \text{ is } A \text{ then } y \text{ is } B$$

where A and B are the fuzzy sets:

$$A = 0.33/6 + 0.67/7 + 1.00/8 + 0.67/9 + 0.33/10$$
$$B = 0.50/1 + 1.00/3 + 0.50/5$$

The implication relation of the rule is modeled through the Larsen Product implication operator given in Problem 4.5. If a fuzzy number $x = A'$ is a premise, use *generalized modus ponens* to infer a fuzzy number $y = B'$ as the consequent. A' is defined by

$$A' = 0.5/5 + 1.00/6 + 0.67/7 + 0.33/8$$

Use max-min composition with Larsen Product implication.

Problem 4.8. The fitness function is the most important quantity in the use of genetic algorithms. Discuss at least two methods used for determining an appropriate fitness function. How could neural networks be used in selecting the fitness function.

References

Gao, Y., *et al.*, "*Retrieval-Augmented Generation for Large Language Models: A Survey*," arXiv:2312.10997, arXiv, 2024.

Goldberg, D.E., *Genetic Algorithms in Search, Optimization, and Machine Learning*, Addison-Wesley, Reading, MA, 1989.

Holland, J.H., "Genetic algorithms," *Scientific America*, 267, 66–72, 1992.

OpenAI *et al.*, *"GPT-4 Technical Report,"* arXiv:2303.08774, arXiv, 2024.

Pantopoulou, S., Ankel, V., Weathered, M.T., Lisowski, D.D., Cilliers, A., Tsoukalas, L.H., and Heifetz, A., "Monitoring of temperature measurements for different flow regimes in water and galinstan with long short-term memory networks and transfer learning of sensors," *Computation*, 10, 108, 2022a.

Pantopoulou, S., Cilliers, A., Tsoukalas, L.H., and Heifetz, A., "Transfer learning long short-term memory (TL-LSTM) network for molten salt reactor sensor monitoring," *Transactions of the ANS Winter Meeting*, 2022b.

Pantopoulou, S., Pantopoulou, M., Prantikos, K., Heifetz, A., and Tsoukalas, L.H., "Monitoring of flow rate measurements using LSTM autoencoder," *15th International Conference on Information, Intelligence, Systems & Applications (IISA)*, pp. 1–4, Chania Crete, Greece. https://doi.org/10.1109/IISA62523.2024.10786694

Pantopoulou, S., Weathered, M., Lisowski, D., Tsoukalas, L., and Heifetz, A., "Temporal forecasting of distributed temperature sensing in a thermal hydraulic system with machine learning and statistical models," *IEEE Access*, 13, 10252–10264, 2025.

Prantikos, K., Chatzidakis, S., Tsoukalas, L.H., and Heifetz, A., "Physics-informed neural network with transfer learning (TL-PINN) based on domain similarity measure for prediction of nuclear reactor transients," *Nature Scientific Reports*, 13, 16840, 2023.

Prantikos, K., Pallikaras, V., Pantopoulou, M., Tsoukalas, L.H., and Heifetz, A., "Agentic retrieval augmented generation for advanced reactor thermal hydraulic system," *Proceedings of the 2024 American Nuclear Society Winter Meeting*, Orlando, FL, 2024.

Tsoukalas, L.H., *Fuzzy Logic: Applications in Artificial Intelligence, Big Data, and Machine Learning*, McGraw Hill, New York, NY, 2023.

Tsoukalas, L.H. and Uhrig, R.E., *Fuzzy and Neural Approaches in Engineering*. Wiley, New York, 1997.

Uhrig, R.E. and Tsoukalas, L.H., "Soft computing technologies in nuclear engineering applications," *Progress in Nuclear Energy*, 34(1), 13–75, Elsevier Science, 1999.

Vaswani, A., *et al.*, *"Attention Is All You Need,"* arXiv:1706.03762, arXiv, 2017.

Chapter 5

The AI-Energy Nexus

5.1 Gardening at the Edge of the Grid

The transformative promise of Artificial Intelligence to reshape society is paradoxically constrained by its gargantuan appetite for energy. Training colossal large language models and operating vast data centers are creating unprecedented demands on our power infrastructure. In this evolving landscape, securing a reliable and affordable electricity grid becomes paramount, and ironically, AI itself can be a critical part of the solution. This is particularly true at the grid-edge, where the AI-Energy Nexus is rapidly expanding to integrate a growing number of distributed energy resources (DERs) alongside sophisticated sensing and computational capabilities. The inherent complexity and dynamic connectivity of the grid-edge demand an AI-based autonomy that far surpasses conventional automation. While electrical power travels at the speed of light, AI can enable stability and resilience through strategic temporal and spatial partitioning, coupled with trustworthy, anticipatory insights into future demand and supply. We term each such autonomous segment a Local Area Grid (LAG), envisioning it as an electrically interconnected "island" capable of achieving security and performance through localized actions. Thus, AI at the grid-edge can reduce complexity and manage connectivity risks, ultimately boosting overall grid resilience. To achieve this vision, we adopt a "gardening" metaphor, where optimizations akin to "pruning" within each LAG can foster a healthy, self-healing, and remarkably resilient electric power grid.

The power of this AI-Energy Nexus hinges on an increasingly capable infrastructure for inferencing, where AI agents continuously develop scenarios for planning, operations, maintenance, optimization, and management of the grid. Distributed Energy Resources (DERs) are key players here; these include, but are not limited to, wind, solar, bioenergy, geothermal, small hydro, micro-reactors, grid batteries, and electric vehicles capable of injecting power back into the grid. AI agents can orchestrate these diverse resources to manage complexity and connectivity risks, enhancing overall grid resilience through anticipatory evaluations of interconnection requests and trustworthy impact analyses that pre-emptively formulate mitigation strategies. Crucially, *deep machine learning* can enable the grid-edge to become truly self-healing via real-time anomaly detection and swift corrective actions that drastically minimize downtime. Beyond reactive measures, self-learning AI agents can implement proactive control strategies, coordinating numerous grid-edge resources for tasks like optimizing DER dispatch, managing demand-side responses, and meticulously maintaining security and privacy across the network.

Figure 5.1 shows schematically the power grid. At the top we have large generators while at the bottom, the *grid-edge*, we have "loads" co-located with small generators including solar, wind and grid batteries. The edge is where distributed generation and consumption come together in a digitized space with using ever more data fed into AI agents with deep machine learning capabilities.

AI agents can model temporal patterns of production, transmission, distribution, and consumption. Several factors play a role here. The patterns depend on the type of customer, e.g., industrial, commercial, or residential; on season; on day of the week; and, on economic and social influences. It is difficult to unearth all these factors at the granularity of the grid-edge without the help of AI. Valuable information can be found in histories, especially recent data that can help agents extrapolate future values including future demand. For example, cold and hot weather both increase the demand of residential customers. High temperature may also decrease the efficiency of some industry utilities and hence increase their electric demand.

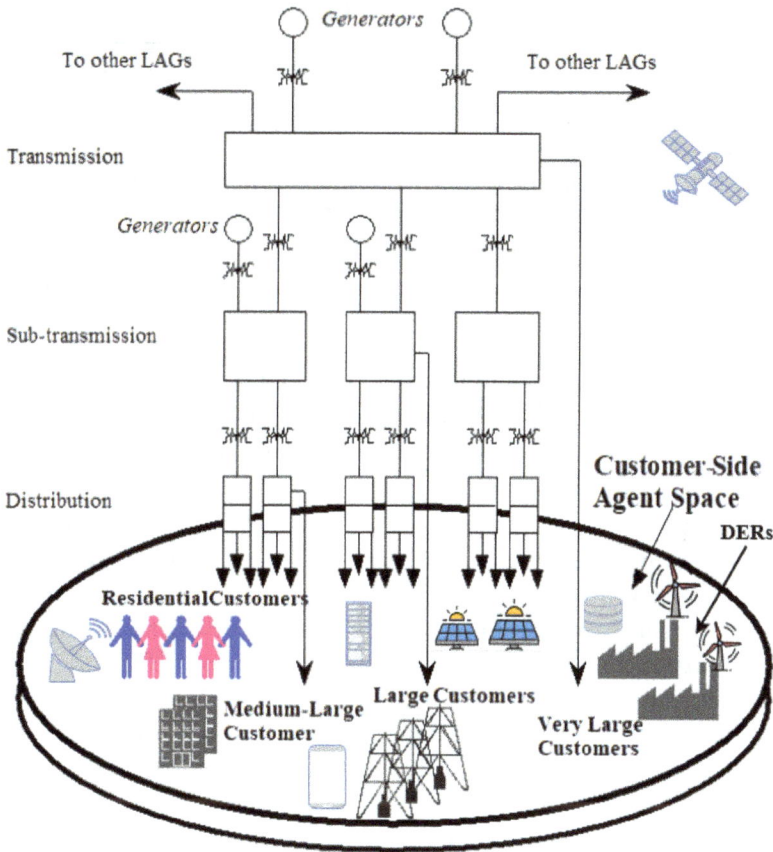

Figure 5.1. Schematic of the power grid with large generators (top) sending power to Local Area Grids (LAGs) through the transmission and distribution networks. LAGs exist at the edge of the grid, or *grid-edge* (bottom), where customers and distributed energy resources (DERs) co-exist. An ever-growing volume of data is collected, and AI multi-agent systems (Agent Space in the schematic) are domiciled at the grid-edge. They use data autonomously, to ensure reliability on the face of growing risks from connectivity and complexity.

Such empirical observations are summarized in Table 5.1. Periodicities based on empirical observations lead to the following heuristics:

- The importance of weather information varies with different customers (Table 5.1).

Table 5.1. Importance of influencing factors.

Customer Type	Recent Load	Weather	Weekend or not
Large Industry	Very Important	Not Important	Very Important
Small Industry	Very Important	Not Important	Very Important
Residential	Important	Very Important	Not Important

- The optimal time window for analysis and daily forecasts is 7 days (a week).
- The optimal size for the analysis window for hourly load prediction is around 20 hours. The most important data includes the same time load of yesterday, the day before yesterday, and so on.

Such empirical observations can be automatically adjusted every day, based on information and results that decide what factors should be included in input and what kind of AI tool may be needed for forecasting future values. They can be learned by agents using deep machine learning such as long short-term memory (LSTMs) neural networks and transformers.

How do we manage all these factors of growing size and complexity at the edge of the grid? For one thing, if complexity exceeds a certain threshold, the edge of the grid may become more fragile, more likely to suffer brownouts and blackouts.[1] In addition, growing connectivity may generate cascading failures, a kind of domino effect where a local anomaly at the edge, if not addressed in a timely fashion, may inadvertently cascade to a global failure.

To answer the dilemmas emerging from ever growing complexity and connectivity we consider the optimization metaphor of *pruning a garden*. This is a metaphor that illustrates the process of refining and improving a system by removing unnecessary or detrimental elements to allow the essential parts to flourish. Pruning strategies have

---

[1]On April 28, 2025, an unprecedented blackout shocked the Iberian Peninsula. The growing complexity at the grid-edge infused with overbearing solar and wind DERs, appears to have been a factor in causing a frequency collapse that led to a catastrophic outage of the national grids of Spain, Portugal and parts of France and Morocco.

Figure 5.2. A garden consists of many different species of plants that can be very heterogenous and if left unmanaged, some species (typically weeds) may take over and may also spill over to neighboring gardens. Hence there is a need for pruning to protect a garden and achieve optimal yields, important functionalities, and desirable esthetics.

been very successful in managing various assets in networks (Mocanu *et al.*, 2018).

As seen in Figure 5.2, a garden may consist of diverse botanical species competing for nutrients, sunlight, and garden space. Plants that are heterogeneous may try to grow against each other with the possibility of substantial *spillover effects* of species into the areas of other species, if unkept. A gardener must coordinate plant growth, often using pruning, so that the overall garden is healthy and various species thrive in harmony with each other and not at the expense of each other. The *Gardener's Dilemma* can be stated as the following question:

How can a gardener find the best combination of pruned and un-pruned plants that will produce the best garden?

The answer in gardening is surprisingly simple.

Partition the garden into patches and answer the Gardener's Dilemma (i.e., optimize) within each patch.

Metaphorically, the garden represents the grid-edge. Garden species such as flowers, weeds, plants, and trees stand for wind, solar, grid batteries, EVs, smart meters, and a growing number of sensors and smart loads responding autonomously to data. Pruning removes dead or diseased branches. In the grid-edge this represents identifying and eliminating inefficient, outdated, or harmful elements including resources destabilizing the system, insufficient storage, congestion, or surges in generation or demand due to connectivity. In general, the grid-edge can become overcrowded like a garden. Too many plants competing for light, water, and nutrients may not be able to thrive. Grid-edge overcrowding may reduce reliability, hence that need to adjust parameters or restructure the system to enhance its performance by pruning non-essential tasks, distracting features, or unproductive efforts that divert resources from core objectives.

The overarching goal of pruning is to encourage valuable outputs and maximize the desired outcomes by eliminating inefficiencies and waste within the system. Effective pruning requires knowledge of different plant types and their specific needs. Similarly, successful optimization requires expertise in the domain and the optimization techniques being applied. If performed correctly, the removal of unnecessary complexity can increase the resilience and lead to a healthier grid. The gardening pruning metaphor highlights that optimization is not just about adding more; it's often about strategically removing the excess to allow the core, valuable components to thrive and achieve the desired results more effectively.

How do we partition the grid-edge into LAGs dynamically? One possibility is to use the concept of *congruence* and define LAGs so that they maintain approximately *medium* congruence (~0.5). Congruence can be effectively measured by the degree of overlap between patches (desirable regions) and spillover sets (unintended consequences), with a value ranging from 0 to 1. A higher value indicates a more congruent partition.

The gardening-pruning metaphor describes a process of iterative refinement or optimization, where patches represent desirable LAG partitions and spillover represent the influence or impact of pruning on neighboring patches. Pruning one area might inadvertently affect neighboring areas. The spillover set captures these indirect effects, and congruence measures the degree to which "pruning" (removing) aligns with enhancing overall performance at the grid-edge. In this sense, "islanding" (the isolation of a patch) may be viewed as the outcome of strategic pruning guided by congruence metrics.

When congruence is high (e.g., close to 1), spillover effects are minimal. When we prune, we primarily affect the unwanted areas and have little negative impact on the areas we want to keep. The pruning action is "congruent" with the goal of preserving patch performance. Low congruence (closer to 0) implies that the spillover set significantly overlaps with the desirable patches. When you prune, you inadvertently damage or remove parts of the areas you want to keep. Thus, the pruning action is *not congruent* with the goal of preserving performance within a given patch.

In Figure 5.3, if we consider the patches (P) and the spillover set (S) as sets within the overall search space. We can define a measure of congruence (C) as

$$C \equiv 1 - \frac{|P \cap S|}{|P|} \tag{5.1}$$

where $|P \cap S|$ is the size (e.g., area, volume, number of elements) of the intersection between a patch and its spillover, and $|P|$ is the size of the desirable patches. Equation (5.1) gives us a value between 0 and 1, with the following interpretation:

If $|P \cap S| = 0$, there is no overlap and Eq. (5.1) gives $C = 1 - \frac{|P \cap S|}{|P|} = 1 - 0 = 1$,

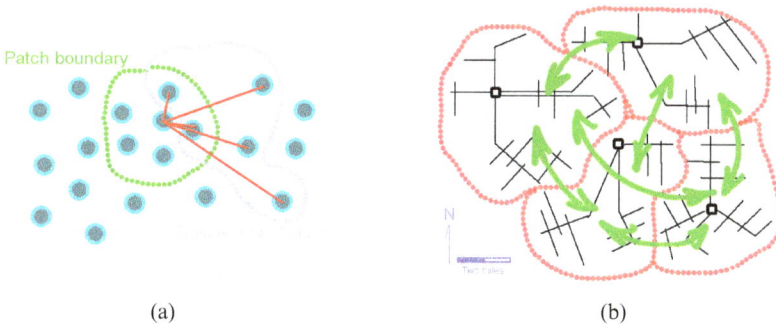

(a)   (b)

Figure 5.3. (a) Garden partitioned into patches – gardener optimizes within each patch. (b) Grid-edge partitioned into LAGs (patches), optimize within each LAG using the anticipatory power of AI. Anticipation reduces unneeded complexity and reduces entropy production; which may enhance grid-edge resilience.

which is the case of perfect dissimilarity. It is important to note that this index is often used to measure dissimilarity or distance between two sets, and it ranges from 0 to 1. When $C = 0, S$ is identical to P (i.e., $|P \cap S| = |P| \, |P \cap S| = |P| \, |P \cap S| = |P|$). At the opposite end when $C = 1$ means that S has no overlap with P (i.e., $|P \cap S| = 0 |P \cap S| = 0 |P \cap S| = 0$).

This formula is useful when you care about the relative overlap of S within P, as opposed to the traditional Congruence Index that looks at both P and its complement.

On the other hand, if $|P \cap S| = |P|$, we have complete overlap and Eq. (5.1) gives $C = 1 - \frac{|P \cap S|}{|P|} = 1 - 1 = 0$, this is a case of no congruence.

Other measures of overlap could be used, such as the *Jaccard index*,

$$J(P, S) \equiv \frac{|P \cup S|}{|P \cap S|} \tag{5.2}$$

Defining congruence as in Eq. (5.2), congruence is related to the *Jaccard index*, a lower Jaccard index indicates low congruence (less relative overlap). The Jaccard Index ranges from 0 (no common elements) to 1 (identical sets).

A highly congruent pruning strategy effectively eliminates undesirable regions of the search space while minimally impacting the patches themselves. Dedicated agents can learn and use the congruence C between the effects of pruning and the preservation of desired solutions can help guide the optimization process and avoid detrimental actions.

In a garden, pruning is optimization applied per patch. In the grid-edge, LAGs are patches and optimization take place within each LAG. With the help of AI resources and functionalities, variables and parameters can be optimized and adjusted proactively within each LAG.

Distributed energy resources in a LAG can be viewed as branches and foliage in a garden. Just as excessive foliage blocks light and air, excessive DERs can create congestion and instability. In this case, pruning is done by implementing smart inverters and control systems to regulate output, curtain excess generation when appropriate, and schedule charging/discharging of batteries accordingly.

Similarly, like a branch not contributing to a plant's yield, under-utilized battery storage may not provide its full value. Developing intelligent algorithms that optimize charging and discharging schedules can be based on price signals, grid conditions, and local energy needs. This "trains" the storage to be more productive. Like vines tangling and hindering growth, unmanaged EV charging can create peak demand surges. Implementing smart charging programs may incentivize off-peak charging, respond to grid signals, and potentially provide vehicle-to-grid (V2G) services. This "guides" the charging behavior and improves LAG health.

Like diseased branches, old transformers or inefficient grid components hinder overall performance. Identifying and replacing outdated equipment with more efficient and intelligent alternatives, "removes" unproductive elements. In addition, like a plant with poor internal transport, a grid edge without robust communication and control lacks coordination. Investing in advanced metering infrastructure (AMI), communication networks, and edge computing to enable real-time monitoring and control. This "improves the internal flow" of information and enables a SCADA at the grid-edge.

Like excessive shade hindering growth, too much raw data without proper analysis may add little value. Hence the "brevity of data," becomes very important for the overall gardening strategy. AI enabled analytics extract meaningful insights from grid-edge data, with better forecasting, control decisions, and anomaly detection. It is analogous to "filtering the light" to provide essential nourishment. Like a plant struggling to absorb nutrients due to soil imbalances, isolated data silos may hinder effective optimization. Developing standardized data formats and communication protocols to enable seamless data sharing and interoperability between different grid-edge devices and systems, may "improve the soil" for better resource absorption.

Undeveloped or inefficient grid services, like buds that don't fully bloom or fruits that don't ripen, might not be fully realized and the pruning in this case may consist of developing and implementing market mechanisms and control strategies that incentivize and enable the efficient provision of grid services, thus "nurturing" development of valuable outcomes.

Lack of customer engagement is like a plant not being properly cared for; a lack of customer participation, for instance, hinders the potential of demand-side management. Pruning corresponds to designing user-friendly interfaces, providing clear incentives, and educating consumers about the benefits of participating in grid-edge programs. This "cultivates" greater engagement and participation.

The gardening pruning metaphor provides a valuable framework for understanding grid-edge optimization as a process of strategically managing and refining the interconnected components at the edge of the grid to foster a more resilient, efficient, and sustainable energy system. It emphasizes the need for careful assessment, targeted interventions, and a focus on the overall health and productivity of the grid-edge ecosystem.

Pruning is related to the management of the second fundamental principle of energy transitions, i.e., *entropy production*. As discussed in Chapter 1 (driving principle 2), the connection between *pruning* and *entropy production* lies in the idea of optimizing a system for resilience by reducing unnecessary complexity.

We recall that entropy is associated with disorder, randomness, and the dissipation of useful energy into less useful forms like heat. Systems naturally tend towards higher entropy over time (Second Law of Thermodynamics). Energy systems like organisms can maintain a state of relatively low internal entropy by constantly exchanging energy and matter with their surroundings, increasing the entropy of their environment in the process. In the gardening metaphor, pruning removes unnecessary branches, leaves, or buds. These parts might consume resources (energy, nutrients) without contributing optimally to the plant's overall health and desired output (e.g., fruit, flowers). Similarly, in a complex system, unnecessary components, redundant processes, or inefficient pathways lead to a higher rate of energy dissipation and growing entropy production than a mere streamlined and optimized patch. Pruning directs resources (energy, matter) towards the essential parts of the system. By removing elements that don't contribute effectively, the remaining components can function more efficiently, potentially reducing wasted energy and the associated entropy production. Pruning uncoordinated DERs or inefficient energy usage patterns through AI can optimize energy flows, reduce transmission losses (a form of energy dissipation), and lead to a more stable and efficient grid. Pruning can simplify complex processes by eliminating redundant steps or bottlenecks. This can lead to faster and more efficient operation, reducing the energy required to achieve a specific outcome and thus potentially lowering the rate of entropy production associated with that process.

Similarly, in machine learning, pruning unnecessary connections or parameters in a neural network reduces computational complexity, leading to faster inference with potentially lower energy consumption per prediction. This can be seen as a form of reducing the "disorder" of the network and making its operation more directed. By removing modeling that hinders performance or introduce instability, pruning helps maintain the overall order and function of the system. This sustained order, while requiring energy input, can be achieved with less energy expenditure in a well-pruned system compared to an overgrown and inefficient one. In biological systems, processes like apoptosis (programmed cell death) can be seen as a form of "pruning" unnecessary or damaged cells, maintaining the overall health and order of the organism. This directed removal contributes to the organism's efficient functioning and survival, influencing its overall energy balance and interaction with the environment in terms of entropy production. Pruning helps focus the system on its primary goals. By eliminating distractions and less important functions, the system can dedicate its energy towards producing the desired outputs more efficiently, while slowing down the entropy production resulting by the pursuit of those goals.

In a garden, the removal of invasive species can lead to a more balanced and efficient flow of energy and resources, and a reduction in the overall entropy production associated with inter-species competition. It's crucial to remember that even with internal optimization, many engineered systems increase the entropy of their surroundings. Pruning aims to make this process more efficient and directed towards maintaining system functionality. Pruning can sometimes involve trade-offs. For example, aggressively pruning a garden might reduce its ability to adapt to future changes. Similarly, over-pruning a technological system might increase its fragility.

The relationship between pruning and entropy production lies in the idea that strategic removal of unnecessary complexity and inefficient pathways can lead to a system that operates more efficiently, potentially reducing the rate of energy dissipation and entropy production associated with its function or directing that dissipation in a more controlled manner to maintain order and achieve desired outcomes. The gardening pruning metaphor provides an intuitive way to understand how eliminating the superfluous can contribute to a more streamlined grid, with reduced rates of entropy production. In essence, pruning helps create a more streamlined and efficient system. However, it is important to keep in mind that even well-optimized systems will still increase the entropy of their surroundings.

On the physical side of the grid-edge, there are important parameters that can be monitored by AI to offer decision support to a Grid-Edge Manager (G-E Manager) – that is a software agent endowed with deep machine learning capabilities and capable of managing the overall system, autonomously. Since electric power travels with the speed of light and no autonomous management system operates at the speed of light, we follow the planning conventions of the electric power industry to divide time into intervals Δt and plan/manage on a per Δt basis. Δt is typically 1 min, 15 min, 30 min, or an hour, day, year, depending on the specifics of an area of the grid-edge and the actions that can to be taken.

For instance, consider average demand, D. As seen in Figure 5.4(a), D is the average energy consumed over an interval Δt. It is defined on a per interval basis and is indicated as the flat line segment in each interval. A sequence of these values (horizontal line segments) gives us a pattern of consumption of energy. Any electricity consuming device like an appliance, a residence, or a hyper data center, produces its own pattern and from the pattern we extract interesting features to or new quantities such as the Peak Demand $D_{\max}$ which is the highest D in all intervals within another time window, that of the period T. It should be noted here, that these temporal partitions and scales have a structure (e.g., Δt's contained within T), and the structure has enormous information value to guide and steer the learning processes that AI can coordinate autonomously at the grid-edge.

Loads at the grid-edge are stochastic. In addition, sources like wind and solar are also (non-dispatchable). Hence temporal partitions in time intervals, $\Delta t \leq 1$ min may be considered. But a coarser temporal granularity may be adequate for autonomous grid-edge management. Figure 5.4(a) illustrates the concept. A time period of interest typically starts at midnight (arbitrarily) and when $t = T$ is reached, we have covered several time intervals Δt. Within each

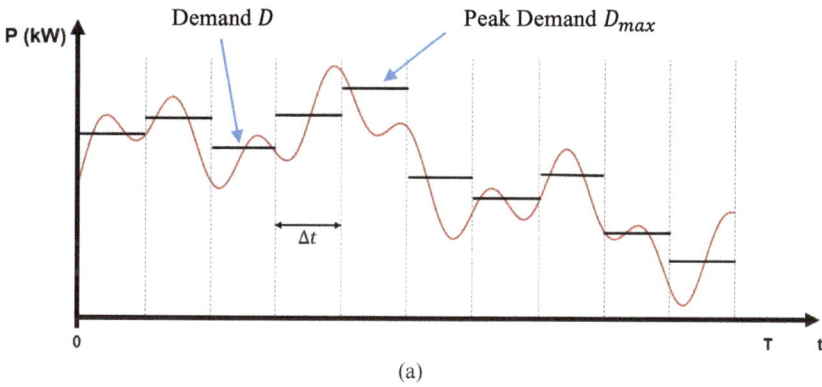

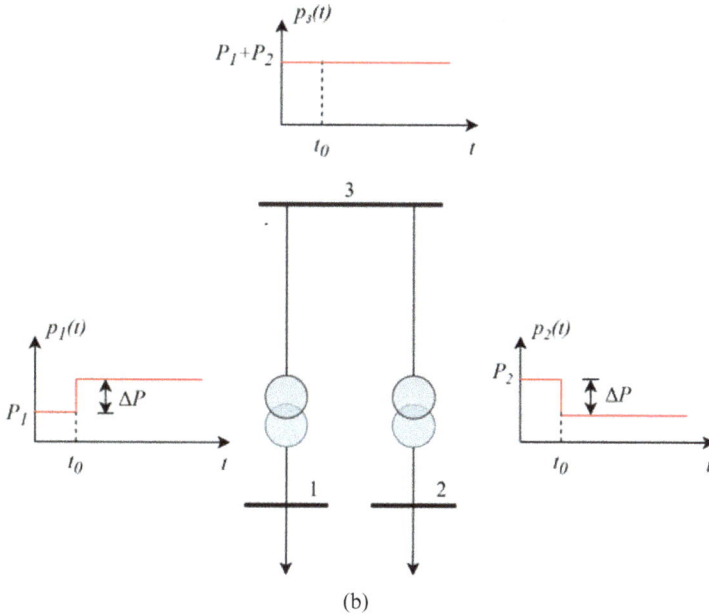

Figure 5.4. (a) A temporal partition in time intervals Δt where within each interval and over a period T, variables, factors and parameters are defined to manage the grid-edge. (b) Foremost goal is matching supply to demand. These are useful for AI to move the grid-edge from automation to autonomy.

interval we can define variables, factors and parameters which are useful for AI agents for the purpose of matching patterns of demand to the power available in a bus, as illustrated in Figure 5.4(b) where the flat profile of the power available at a bus, can be adding the demand patterns of the 2 loads which are complementary in a temporal sense, when the one goes down the other goes up and ideally the two together match the power supplied at the bus, thus maximizing the efficiency of the grid-edge through synchronization between supply and demand for optimal utilization of the available energy resource. Some of the variables that may be used for AI autonomous coordination are listed next.

Average Demand: The average demand of a load (customer) during a specific time interval $t_0 < t < t_0 + \Delta t$ is given by

$$D = P_{av} = \frac{1}{\Delta t} \int_{t_0}^{t_0+\Delta t} p(t)dt \tag{5.3}$$

or,

$$D = \frac{E_{\Delta t}}{\Delta t} \tag{5.4}$$

where

$p(t)$ = power at time t, in units of kW for residential and small commercial and MW for industrial customers, and,
$E_{\Delta t}$ = energy during the time interval Δt.

Peak Demand: This is the maximum demand D_{max} of an individual load, or a cluster of loads in a LAG or many LAGs, during a time window from $t = 0$ to $t = T$ (typically from midnight to time T where often $T = 24$ hours). It is defined as

$$D_{max} = \max_{t=[0,T]} D(t) \tag{5.5}$$

Average Demand over T: The average demand over a period of time T is defined as the average of the average demands in each interval of time up to the time T, that is, it is defined as

$$D_{av} = \frac{1}{T} \sum_{i=1}^{n} D_i \Delta T_i \tag{5.6}$$

Diversified and Coincident Demand: An important set of variables in the grid-edge are *diversified demand* and *coincident demand* of k loads. The first is their sum over a period of time. The second is summing them all together instantaneously. Let a cluster of k loads in a LAG. The degree to which they are coinciding in a temporal sense is given by the sequence of average demand, that is

$$D_{d,i}, \quad i = 1, \ldots, n \tag{5.7}$$

Diversified demand is the sum of the demands produced by a group of loads over a particular period of time. It accounts for the fact that different loads peak at different times, leading to a lower total demand than the sum of individual peak demands.

On the other hand, *coincident demand* is the sum of all demands at any instant of time (not a period). For this reason, it is also called *simultaneous demand*. It represents the total load seen at a given moment, such as at a substation or feeder.

Maximum Diversified Demand: The *maximum diversified demand* is defined as

$$D_g = \max_{i=1,\ldots,k} \{D_{d,i}\} \tag{5.8}$$

The *maximum noncoincident demand* of a group of k loads is defined as the sum of the maximum values of the individual demands in the cluster. Maximum non coincident demand is an instantaneous quantity independent of time intervals,

$$D_{nc} = \sum_{i=1}^{k} \{D_{\max,i}\} \tag{5.9}$$

Demand Factor: The demand factor of a load is defined as the ratio of maximum demand to the rated (nominal) power of the load, i.e.,

$$DF = \frac{D_{\max}}{P_{\text{nom}}} \tag{5.10}$$

Utilization Factor: The utilization factor F_u of a cluster of k loads in a LAG is defined as the ratio of maximum coincident demand of the cluster to the rated capacity (or nominal capacity) of the cluster. Hence,

$$F_u = \frac{D_g}{P_{\text{nom}}} = \frac{D_g}{\sum_{i=1}^{k} P_{\text{nom},i}} \tag{5.11}$$

where $P_{\text{nom},i}$ = nominal (rated) power capacity of any load i.

The nominal (rated) power capacity of a load is "named" (hence the term "nominal") as a technical specification given by the manufacturer or designer of an electrical machine. It is based on the power level that the machine can utilize without violating thermal limits (i.e., overheating), or the maximum voltage drop. In addition to loads, a rated capacity can characterize all power consuming devices and systems in a LAG including transformers and wires (power lines).

Load Factor: The load factor of a customer is defined as the ratio of the average load over the peak load in a given period of time, that is,

$$F_{LD} \equiv \frac{D_{av}}{D_{max}} \times 100 \qquad (5.12)$$

The load factor is a measure of how efficiently electric energy is being used. For a group of loads (cluster) we can define the load factor as

$$F_{LD} \equiv \frac{D_{d,av}}{D_g} \times 100 \qquad (5.13)$$

A higher load factor indicates a more consistent and efficient use of electrical power, reducing strain on the grid and lowering costs. Conversely, a lower load factor suggests sporadic or inefficient usage, leading to higher peak demand costs.

Coincident Factor: This is a measure of how k loads may have similar versus diverse synchronization. It is defined as the ratio of the coincident peak demand to the sum of individual peak demands, i.e.,

$$F_c = \frac{D_g}{D_{nc}} = \frac{D_g}{\sum_{i=1}^{k} D_{max,i}} \qquad (5.14)$$

A higher coincident factor means that loads peak at the same time, requiring more capacity while a lower coincident factor indicates that loads peak at different times, reducing overall demand on the system.

Diversity Factor: This is a measure of how different loads peak at different times, reducing the total demand on the system. It is defined as:

$$F_D \equiv \frac{\text{Sum of Individual Peak Demands}}{\text{Coincident Maximum Demand}}$$

or,

$$F_D \equiv \frac{D_{nc}}{D_g} = \frac{\sum_{i=1}^{k} D_{\max,i}}{D_g} \qquad (5.15)$$

or, using Eq. (5.9), we have

$$F_D = \frac{\sum_{i=1}^{k} P_{\text{nom},i} \times DF_i}{D_g} \qquad (5.16)$$

A higher diversity factor (greater than 1) means that loads peak at different times, allowing the grid edge to have lower capacity requirements and AI to coordinate loads as illustrated in Figure 5.4(b). The net effect of such coordination is to increase grid-edge resilience and reduce costs in the power grid.

Many of the factors have been historically used for planning grid expansions and installations. Yet with AI's learning capacities brought at the grid-edge, the same factors estimated or predicted on a per interval Δt or a period T basis, can be used for the autonomous management of the grid. Of particular interest are demand analytics that break down demand on weather or seasonal components in addition to the stochastic aspects, e.g.,

$$D(t) = D_{nw}(t) + D_w(t) + D_s(t) \qquad (5.17)$$

where

$D_{nw}(t)$ = weather independent demand
$D_w(t)$ = weather dependent demand
$D_s(t)$ = stochastic demand

Deterministic forecasts of demand (and other grid-edge variables) include,

$$\text{Linear regression,} \qquad y = at + b$$
$$\text{Polynomial regression,} \qquad y = a + bt + ct^2$$
$$\text{Exponential regression,} \qquad y = a(1 + b)^t$$

Various statistical techniques have been used for non-deterministic predictions. As developed in the following sections, exponential growth in grid-edge data enables deep machine learning methods for grid agents with superior prediction capabilities. Not only at steady state or equilibrium conditions, but also for transient or departure from equilibrium conditions, especially when the grid-edge has growing stochasticity due to injection of power from wind and solar (non-dispatchable sources).

In addition to the temporal partitions, spatial partitions of loads need to be taken into account. In a service area **A**, at a given time, we can define demand in geographic terms, that is,

$$D(x,y) = \frac{dS}{dA} \tag{5.18}$$

where

A is the service area in units of m^2, and,
S is the total load in the service area in units of kilovolt-Ampere (KVA).

Hence, the units of $D(x,y)$ will be in KVA/m^2. Spatial models of load density can be used to estimate dynamically voltage drops and power losses per LAG and determine the voltage needed to achieve balanced conditions in the grid-edge.

Example 5.1. Diversity factor and Coincident Demand.

If in a LAG, individual peak demands sum to 500 MW and the coincident maximum demand is 400 MW, what is the diversity factor F_D and how can be optimized.

Solution: Using the diversity factor F_D of Eq. (5.15)

$$F_D = \frac{500 \text{ MW}}{400 \text{ MW}} = 1.25$$

The diversity factor can be optimized by implementing load scheduling and demand response strategies such as shifting non-essential loads to off-peak hours using anticipatory smart scheduling or pricing signals that incentivize load adjustments.

Using battery energy storage systems (BESS) to smooth out demand fluctuations by storing excess energy during low-demand periods and dispatch it during peak times. Flexible HVAC and cooling systems can be optimized to prevent unnecessary spikes.

AI-driven load forecasting can by used to balance demands across different loads.

Example 5.2. Load factor calculations for a hyper data center.

A hyper data center (large data center campus) operates for 24 hours a day and has total energy consumption per day equal to 19,200 MWh (assuming an average demand of 800 MWh usage over a 24 hours). The peak power demand (highest instantaneous power usage) of 1,000 MW (1 GW),

(a) What is the *load factor* of the hyper data center, and
(b) Discuss how the *load factor* be optimized.

Solution:

(a) First, we calculate the average demand over the 24 hour period $(T = 24 \text{ h})$ using Eq. (5.6), i.e.,

$$D_{av} = \frac{\text{Total Energy Consumption over } T}{T}$$

$$D_{av} = \frac{19,200 \text{ MWh}}{24 \text{ hours}} = 800 \text{ MW}$$

Next, we calculate the *demand factor* (or *load factor*) according to (5.10)

$$DF = \frac{D_{av}}{P_{nom}} \times 100\%$$

or,

$$DF = \frac{800 \text{ MW}}{1000 \text{ MW}} \times 100\% = 80\%$$

It should be noted that this is a rather high demand factor (close to 100%) which indicates relatively stable that is constant power usage as is fairly typical of data centers due to continuous machine learning computing loads. Conversely, if the demand drops significantly during off-peak hours, it may result in under-utilized power system and thus higher electricity costs.

(b) Optimizing the load factor DF *means* increasing it to reach as close as possible to 100%. A DF approaching 100% is crucial for improving energy efficiency and reducing costs, especially in a large-scale facility such as a hyper data center.

To achieve higher *DFs*, we can implement a plethora of options that may fit the needs and conditions of the hyper data center. For instance, individual loads within the facility can be shifted by scheduling non-essential tasks (e.g., cooling or back up system testing) for off-peak hours. To achieve this, predictive analytics can be used to schedule workloads in analogous ways to what is shown in Figure 5.4(b); in addition scheduling to match demand(s) to supply can prevent unnecessary power spikes.

Grid-scale battery storage can be used to smooth out demand fluctuations and store excess energy during low-demand periods and dispatch it locally (within the LAG) during peak-demand.

In the future, additional options may include micro reactors and small modular reactors on-site as well as DERs such as solar and wind with battery storage and gas turbines. It should be noted that micro reactors and SMRs are not as flexible in ramping up-and-down to follow fluctuating demand. Anticipatory autonomous control (as described in Section 5.6) may enable such flexible modes of operations while improving overall operations and maintenance costs. AI-driven cooling optimization, thermal energy storage, load aggregation and intelligent load balancing are some additional options to consider in order to make the *DF* of the hyper data center approach 100%.

Example 5.3. Predicting peak demand at hyper data centers.

A company operating several hyper data centers over the past 10 years has peak demand shown in GW as shown in the table below. What is the anticipated peak demand at the 11th year?

Year 1	Year 2	Year 3	Year 4	Year 5	Year 6	Year 7	Year 8	Year 9	Year 10
3.56	4.44	5.12	4.91	5.35	5.91	6.32	7.13	8.12	9.44

Solution: To predict the peak demand at the 11th year, we can analyze the trend in the data and apply a suitable forecasting model. Since the peak demand fluctuates slightly but shows an overall upward trajectory, a simple linear regression or exponential smoothing approach may work well.

Let's use a linear trend by fitting the data through a liner least squares regression. In this manner we can approximate a deterministic linear model

$$y = at + b$$

where

t is the independent variable representing the year index (1 to 10)
y is the dependent variable representing peak demand in GW
a is the slope (rate of change per year)
b is the y-intercept

Next, we estimate the slope using the least squares method, that is

$$a = \frac{n \sum_{i=1}^{i=10} y_i t_i - \sum_{i=1}^{i=10} y_i \sum_{i=1}^{i=10} t_i}{n \sum_{i=1}^{i=10} t_i^2 - \left(\sum_{i=1}^{i=10} t_i\right)^2}$$

where $n = 10$ (number of years)

$$\sum_{i=1}^{i=10} t_i = 1 + 2 + 3 + \cdots + 10 = 55$$

$$\sum_{i=1}^{i=10} y_i = 3.56 + 4.44 + \cdots + 9.44 = 61.34$$

$$\sum_{i=1}^{i=10} y_i t_i = (3.56)(1) + (4.44)(2) + \cdots + (9.44)(10) = 389.86$$

$$\sum_{i=1}^{i=10} t_i^2 = 1^2 + 2^2 + 3^2 + \cdots + 10^2 = 385$$

Inserting these values in the equation for a gives

$$a = \frac{10(389.86) - (55)(61.34)}{10(385) - 55^2} \approx 0.636$$

For the y-intercept b, we have

$$b = \frac{\sum_{i=1}^{i=10} y_i - a \sum_{i=1}^{i=10} t_i}{n}$$

$$b = \frac{61.34 - (0.636 \times 55)}{10n} \approx 2.636$$

Thus, for $t_{i=11} = 11$

$$y_{11} = (0.636 \times 11) + 2.636 = 9.636 \text{ GW}$$

Hence, the predicted peak demand for year 11 is approximately 9.64 GW (using a linear regression model).

More refined estimates considering seasonal effects or other patterns can benefit from *polynomial regression* $(y = a + bt + ct^2)$ or exponential regression, that is of the form $y = a(1 + b)^t$. Advanced neural or logical AI predictors can use deep learning to develop models from the data automatically. They include a plethora of neural and fuzzy (logical) tools (Tsoukalas and Uhrig, 1997). More advanced methods include Long Short-Term Memory (LSTM) networks (see Example 5.7) or Transformers (see Example 5.8) or some combination of transformed quantities and neural learning as shown in Section 5.7 (Tsoukalas, 2023; Pantopoulou *et al.*, 2025).

5.2   AI Functions in the Grid

The top objective of the AI-Energy nexus is preventing outages under all possible contingencies. Predictive models with deep learning capabilities could be very useful in managing the AI-Energy nexus to achieve this objective. The models are embodied by agents that benefit from data-driven machine learning capabilities and can transfer what they learned to other agents. We take a closer look at predictions to achieve the top objective, starting with short-term predictions.

In the grid-edge short terms predictions refer to time scales that range from seconds (known as very short-term) to several minutes. Due to the existence of strong temporal correlations, short-term load prediction can be done accurately by neural and deep learning methods with plasticity and further learning capabilities as needed in the future. In addition, some methods can be endowed with transfer learning, which goes past plasticity to enable agents to share learning from different scenarios (Prantikos *et al.*, 2023).

Ideally, the AI-Energy nexus should have self-healing capabilities as seen in Figure 5.5 where five of states as shown. If the NORMAL operating states (frequency and voltage constraints) are violated, the grid may go into ABNORMAL states, where it has limited time to

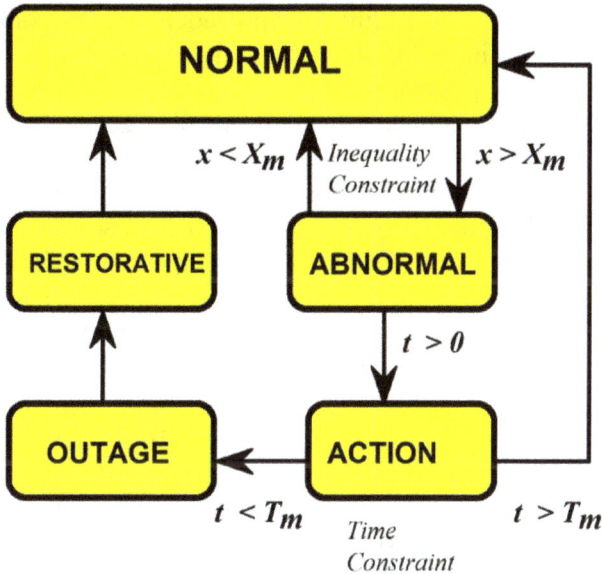

Figure 5.5. A typology of principal grid states, where grid autonomy ensures returning from ABNORMAL to NORMAL without going through the ACTION-OUTAGE-RESTORATIVE path.

achieve corrective actions that bring it back to normal. The typology of grid operating states in this framework is as follows:

Normal State: Critical variables stay within normal limits.
Abnormal State: It is important to relieve the abnormality.
Action State: Actions must be taken.
Outage State: Faulty part(s) removed from service.
Restorative State: Inspections/repairs to reach normal.

It should be noted that time is critical here, with a specific time T_m required to remove an abnormality (Amin, 2013). If the grid can take successful corrective actions on its own before T_m then it is capable of self-healing. If not, the abnormality may evolve into an outage which requires significantly more time (hours or days) to be addressed through actions that eventually will restore operations to normal state. To address the temporal dependence of grid the sensing system generates data and AI generates data-driven models. AI can then use these models to make timely decisions autonomously (e.g., demand side management strategies or small generator or battery dispatch). The grid-edge acquires unprecedented autonomy through self-learning capabilities for stability and performance by anticipating future states and action scenarios.

At the grid-edge, agents can manage locally "generation/demand" realities. Loads that bear digitization can be identified and prioritized in terms of their impact on grid-edge maintaining normal state. For digitized critical loads, historical data can be used for learning as well as forming digital twins that can facilitate transfer learning to other critical loads. Nonetheless, we can have zoning for weather-specific and customer-specific load forecasts. The approach can lead to automated fine-tuned resource optimization and proactive security assessment.

To achieve these results, it is imperative to have reliable prediction of local behaviors. Hence, we need means to characterize the predictability of loads and ensure that AI is capable of automatic decisions that maintain or restore normal states. Consider for example the annual electric power consumption of a large commercial industrial customer (consuming 300–600 MWe) shown in Figure 5.6. The graph shows all hourly variations (since the beginning of May and it covers an entire year). We can see in Figure 5.5, the weekly (weekdays vs weekends) variations in the demand as well as seasonal variations and holiday patterns. A large load like this is expected to be predictable, but its predictability varies with time. In general, large commercial/industrial customers possess predictable consumption patterns. To the contrary, small residential customers are too

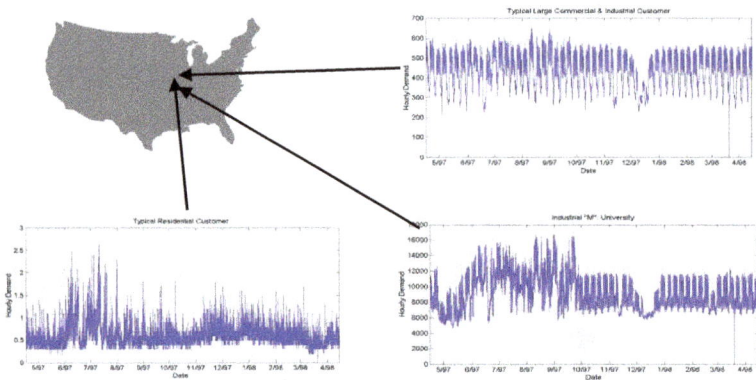

Figure 5.6.   Annual hourly consumption of *Residential, Industrial* and *Commercial* (University) loads.

chaotic to forecast. However, the aggregation of certain number of small customers greatly increases predictability.

How can predictability be dynamically estimated? A multi-scaled analysis tool – wavelets – can be used to perform predictability analysis and in addition, uniquely identify the load. For instance, a large office customer, when off hours, is using energy as if it was a residential customer. We use the concept of phase portraits, to reveal these modes. As seem the phase portraits of Figure 5.7, where the loads of Figure 5.6 are shifter by 6 hours, all loads display some residential load characteristics corresponding to lighting and air conditioning for comfort and safety of humans within a factory or commercial building, but also additional features that uniquely fingerprint a facility. For instance, within 24 hours, there is a cycle of about 6 hours where the mode of energy consumption changes, and different aspects of the mode can reveal if it is energy used to produce things, or, energy used for comfort and security.

To address the question of predictability, we seek help from a wavelet analysis that breaks the load into several detailed analytical components. Wavelets are a multi-resolution analysis technique capable of selecting localized features. If we take the product of detailed

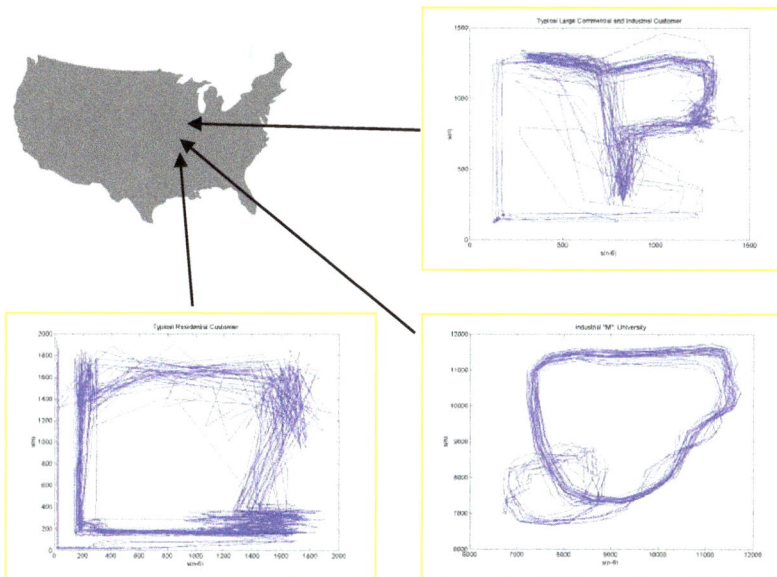

Figure 5.7. Phase diagrams "fingerprint" *Residential, Industrial* and *Commercial* loads.

groups, we find that there is hidden information there. We look deeper into wavelets, but detailed groups do provide a measure of predictability.

The random component in a time series can be discovered automatically by AI. If measurements are drawn from some underlying probability distribution, then we are at a truly random time-series. Agents can characterize the data by statistical moments such as mean, variance, kurtosis, considering short-time scale variations. In addition, AI agents can decide whether the data is not random, but chaotic. For instance, the data can be the outcome of a deterministic processes with nonlinear dynamics. In this case, the AI program can examine Lyapunov exponents to proceed with proper model creation.

The application of feature extraction and decomposition techniques on statistical and machine learning models for load forecasting is well established in the literature. Methods have been developed utilizing the point forecasting metrics of absolute and relative error as well as the comprehensive error metrics of MAE, RMSE and MAPE denoting high fitting ability and error stability. Support Vector Regression (SVR), Deep Neural Network model (DNN) for very short-term load forecasts based on correlation and principal component analysis for feature selection and EMD for pattern extraction. Variational Mode Decomposition (VMD), an autoencoder and an LSTM network for average and peak demand predictions evaluated through MAPE graphs and RMSE for the training and test set. This complex architecture outperformed state-of-the-art demand prediction models denoting the potency of decomposition-based hybrid approaches at the cost of convergence. An ensemble consisting of LSTM members to derive the aggregated predictions have yielded lower error metrics when compared to standalone LSTM, LR and XGBoost approaches, outlining the significance of time-frequency decomposition methods to address load uncertainties and non-linearities.

5.3   Multi-Agent Systems

Multi-agent systems are networks of software entities, called *agents*, that at the grid-edge collaborate to optimize power generation, distribution, and consumption in a harmonious way. Agents monitor, control, and optimize power distribution based on real-time data. They are capable of data-driven learning, including deep machine learning, based on which they can anticipate future grid states. This is important for "gardening." Trustworthy anticipations of the future can empower the grid-edge giving it increased degrees of autonomy, flexibility, and fault-tolerance. For instance, if one agent fails, others maintain grid stability by rerouting resources. Still others may interface with remote substations to ensure continuous power flow. Additional benefits include enhanced efficiency, increased resilience and seamless integration of distributed energy resources with autonomous fault handling capabilities.

Agents act on behalf of "agency" emanating from other agents or humans. These agents have some degree of autonomy and can move around a network to accomplish their defined tasks. They can learn on their own, transfer their learning to other agents while they can be endowed with knowledge of goals, agendas and means to achieve tasks.

Multi-agent systems provide for a modular, extensible approach to problem solving. The agents in a multi-agent system are collections that can dynamically organize themselves into organizations. These agent organizations can be hierarchical, democratic or some other form depending on the task involved. An agent senses the environment and acts on it in pursuit of agendas that may broadly defined teleologically and modified in the fullness of time through machine learning. Agents, often called "intelligent agents," are characterized by autonomy (ability to act independently), social ability (ability to communicate with other agents through an inter-agent communication language), reactivity (ability to react to changes in the environment) and proactivity (ability to influence its environment in anticipation of future changes).

Agents are different from traditional client/server communication (see Figure 5.8) because they can migrate from one point to another taking with them, not just their data state, but also their execution state. As the figure indicates, they take advantage of local resources at each node in the network, making each node, in effect, a server. This is extremely important from the point of view of robustness. Agent technology significantly improves robustness even considering low network bandwidths and unstable connectivity. To develop a framework for management and control of the grid an agent-based

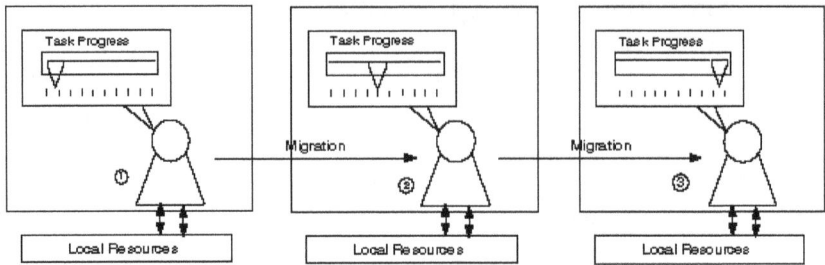

Figure 5.8.   Agents collaborating to complete a process.

framework is appropriate for inter-process communication between the various components.

At the highest level, users can monitor and control agents under their jurisdiction. For instance, a user at the level of a customer can monitor the activity of that specific customer agent while compliant with privacy protection constraints. A user at the level of a utility can monitor the activity of utility-accountable agents as well as that of agents of customers buying power from that utility. At the next level of the hierarchy come the tools for load prediction, security analysis and control strategies like demand side management (DSM) or distributed generation. Agents have access to a plethora of tools including trend-analysis, data mining, deep machine learning, and anticipation, integrated into this level of the hierarchy. A layer, called "Interbase" in Figure 5.9, has agents that communicate with all the databases that are required to perform the calculations in the previous layer (e.g., load prediction).

The actual flow of information among the various entities in Figure 5.5 is determined by a training agency, which combines tools for forecasting like trend analysis, data mining, neural networks etc., on a per-load, or LAG basis. Once the weights and biases are initialized after training, the trained networks are used to predict previously known load values, and these can be compared with the actual values to verify the accuracy of the predictions. Once the accuracy of the predictions is ascertained, the networks are used for predicting variables of interest including, voltage, frequency, demand, congestion, generation etc., several time steps into the future, and based on these predictions, a stability/security analysis is performed with the power-flow solutions and based on this analysis, control actions may be initiated, or "pruning" strategies considered.

Interface		
Load Prediction	**Stability Security**	**DERs DSM Storage**
Interbase		
Multiple Databases *Weather, Load History, Calendar, Financial Information*		

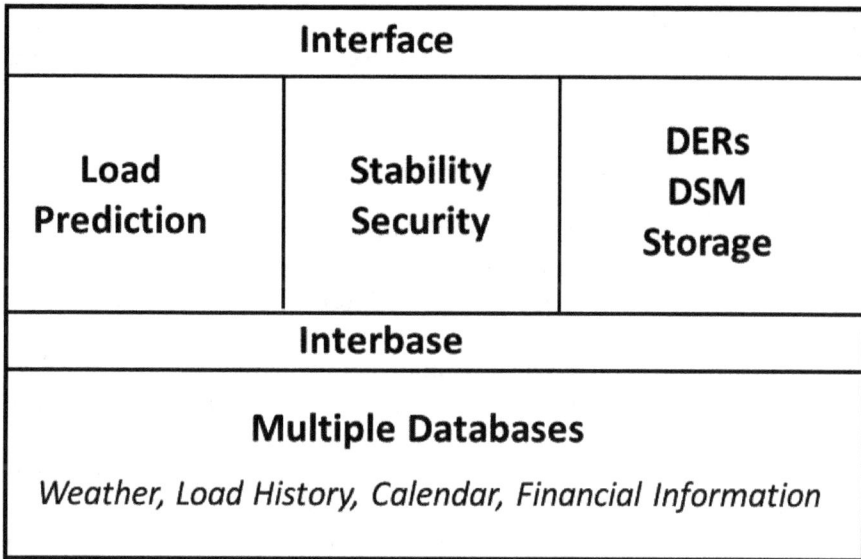

Figure 5.9. Grid-edge agents assume various roles including monitoring and ensuring functionalities such as load prediction, stability/security, generation, and demand side management through appropriate data brevity from databases.

Grid-Edge is the set of interactions emanating from the convergence of the traditional, centralized grid and the evolving, decentralized energy landscape. Components and functionalities can be assigned agents serving them in a dedicated but flexible way, meaning with capabilities for deep learning, transfer learning, explainability and, eventually, measurable trustworthiness. Grid-edge builds new layers of structures (and thus new agencies) on the conventional electricity grid including the decentralized layers of distributed energy resources (DERs), new sensing, and demand-side management. Traditionally, electricity flowed one-way, from large power plants to consumers. Now, with the rise of solar panels, battery storage, and other DERs, energy can flow in both directions. It includes, but is not limited to, rooftop solar panels, electric vehicle charging stations, grid batteries, smart appliances and microgrids. It requires new technologies and strategies to manage the flow of electricity and ensure grid stability while allowing for the convergence of renewable energy, energy storage and small devices.

In a simplified sense, Grid Edge is a system of producers and consumers of electric energy (or groups of consumers). This is a logical organization and need not necessarily reflect geographical proximity between the consumers and producers. The central controlling agency could be a Grid-Edge or G-E Manager. Each producer or consumer agent communicates important data to the G-E Manager; thus, the G-E Manager has complete information about the status of the LAG at any given point of time. To incorporate fault-tolerance into the system, the producer/consumer agents may communicate with a neighboring G-E Manager as well.

A customer agent shown in the figure can represent a large commercial/industrial customer or a group of small commercial or residential customers. Each customer agent has local database and prediction agents to perform load forecasts for that consumer (or group of consumers). Each agent can be represented as in Figure 5.10.

The power meter logs power usage and feeds the data to the local database as well as to the prediction agents. The prediction agents use this current data as well as historical data from the database to make load forecasts for a specified time into the future. These forecasts are put back into the database and communicated to the G-E Manager along with the current load (as measured by the meter). The decision module then makes decisions based on the local forecasts and from the directives, if any, from the G-E Manager. The decisions are then conveyed to the effectors, which execute the final control action. This could be in the form of bringing online captive generation or shedding some local load.

The G-E Manager gets information about the current demand in the G-E, the amount of local generation available and inflows/outflows from neighboring LAGs. Also, the G-E Manager knows all the contracts the consumers have concerning curtailment programs. Hence, with authorization from the power company, the G-E Manager can choose to curtail power for certain customers, if the situation demands such an action. The G-E themselves can be organized hierarchically, with entities at the top having a more abstract view of the system. Tasks like optimal power flow solutions and system wide security analysis can be handled at these higher levels.

In Figure 5.10, there is a substation with three transformers, a few feeders, and some loads (customers). Each physical entity such as a customer (load) or a feeder line, is represented as an agent in the system. The agents are organized hierarchically according to the topology of the power network. The customer agent makes load predictions (based on past load history) and communicates the estimated power

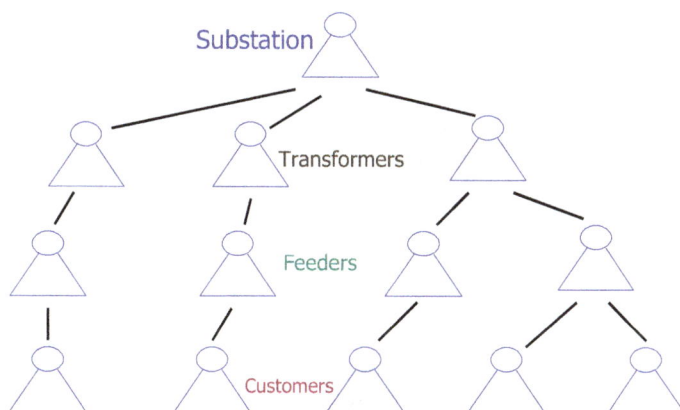

Figure 5.10. Hierarchy of Grid-Edge Management with dedicated agents.

requirement to its feeder agent. The estimated power is also written to a database. The feeder agent sums up the requirements of all the customers beneath it and communicates this to the transformer agent. The feeder agent also checks whether any physical constraints will be violated if this estimated power were to flow through it over the next time step. For example, if satisfying the power requirement will result in overloading of the feeder, then this agent will flag a warning. The transformer agent is similar in function to the feeder agent.

The customer agent consists of feedforward neural networks performing the task of load prediction. The neural network is trained upon arrival at the customer's agency using the load history of the customer stored in a local database. Once it is trained, the agent makes predictions for varying time-steps into the future. All the customer agents on the system are synchronized to work at a specified time at the beginning of each time-step.

The feeder agent waits for the customer agents to make their predictions and then sums up their predicted values. The feeder agents then communicate this value to the transformer agent and sleep till the customer agent is ready with the predictions for the next time step. The transformer agents do the same thing. The substation may sum load predictions from each transformer and inter-agent communications take place in agreed upon protocols.

A central agency, called the *distribution agency* initially spawns all these agents and dispatches them to their respective locations. The registry maintains the location of all the agents in the system. The system can run reliably, if agents do not go out of synchronization due to differing computational loads at each agency or malicious cyber interference for which special agencies may be assigned to protect the multi-agent system (McDonald and Thomas, 2019). Network delays may affect the performance of the feeder and transformer agents since these have to make a remote request for data.

The main goal of G-E Manager is guarding reliability and efficiency within an area without compromising the reliability of neighboring LAGs and by extension the reliability of the entire grid. These goals are achievable through the balance of the power flow in a customer-driven model of the system. To maintain effective supply-demand relations and the needed equilibrium, we need some sort of monitoring and controlling mechanism to watch the power flow through each segment of the grid. Generally, two kinds of control mechanisms exist for such purposes: centralized control and distributed control. Centralized control is not very attractive for complex system modeling. First, it is rather impractical to consolidate all responsibilities to one central agent; it requires huge communication resources and network bandwidth. Second, a central control agent may create a severe performance bottleneck for the entire network of agents. It is intuitively obvious that adding another layer of complexity on the grid (that is agents) is not simplifying or enhancing the reliability of the system. Therefore, a distributed architecture, where responsibilities are spread among a set of agents, offers a successful and reliable solution to a complex system. In a distributed system, each agent is assumed to have some degree of autonomy. There is no single powerful agent with responsibility for managing other agents. Thus, hardware and software requirement can be subsequently lowered (from the point of view of costs) and overall system reliability enhanced. Robustness is an inherent feature of distributed systems. Since the inference amongst different agents has been reduced a minimal level, the corruption of one or several agents may not cause a large-scale negative effect on the grid. An important question concerns the degree of autonomy of individual agent. Should each agent be fully autonomous and thus behaves independently? Alternatively, should some hierarchical control agents be governing a small set of agents? When studying these questions, one finds some interesting possibilities. A series of challenges emerge from the complexity of grid-edge.

- The scale of the problem is at least the same, if not bigger than, what is found when one looks at the Internet alone (a system hosting millions of computers). Any solution to the complexity of the grid should be examined within this scope and should not just work for a small group of entities. Innovations on the grid-edge should be both technically and economically feasible.
- Stability is our major concern for a complex system. This means that faults should be isolated and should not spread within the network.
- The solution should be scalable, which requires that it's open to the dynamical changes of the system: either in size or in structure. Considering trends of growing electrification and power utilities deregulation, this requirement is mandatory.

Intelligent meters will further complicate the existing network, where huge number of meters come into the scope of the problem. To meet the coming challenges, a feasible solution is expected to possess the following properties:

- Each agent should be relatively independent. It only exchanges necessary information with its environment. It is functionally self-contained, which enables it to function with minimal interference from outside.
- Agents themselves should be passive. A passive agent will be disconnected from the network if it malfunctions. This also requires each agent has some sort of self-diagnosis and self-healing capability. Whenever an internal, incurable error is detected, the agent isolates it from the others.
- Protocol-based robustness is preferable than implementation-based robustness. By protocol-based robustness we mean that any exception should be considered and handled explicitly by the protocol.

Examining the abovementioned requirements, we can conclude that only a distributed solution can meet all these challenges.

Within the distributed framework, any basic agent ontology should distinguish between three classes of basic agents (with sub-classes within each class possible): generation agents, transmission/distribution agents, and consumption agents. All agents are functioning independently, and they communicate through messages. Below we list a typology of basic agents.

5.3.1   *Generation agent*

The generation agent is responsible for scheduling the generation of power. The first thing it will do is broadcasting pricing information to all accessible customers and then waiting for consumers' order. Upon receiving all orders, the generation agent sums all requested demand and check if its capacity can meet all the orders. If yes, all orders will be accepted. Otherwise, it can curtail customers' demand by increasing the price.

5.3.2   *Transmission/Distribution agent*

Transmission agents are used to monitor the load status on transmission lines. Each agent is responsible for a segment of transmission line. If an estimated power flow exceeds the limit, the transmission agent will ask customers or downstream transmissions (or distribution) to plan in an anticipatory manner to reduce their future power consumption. If a real-time overflow is detected, the transmission agent will require consumers to reduce power consumption immediately. Otherwise, it will curtail the power in a mandatory fashion. Transmission agents can make unit dispatching easier if a transmission cost is added to the total price by each transmission agent.

5.3.3   *Consumption agent*

The consumption agent resides in every intelligent meter. The major task of a consumption agent is to manage the electricity consumption of the consumer. The agent will first predict how much power it will need shortly. Then it will decide on where to order the cheapest electricity based on the pricing information it has received. In addition, the order is unsuccessful due to the transmission or generation capacity; it then will try to either buy electricity from other companies or curtail its power consumption.

As an illustrative example consider the grid shown in Figure 5.10. Here the grid includes three types of agents: *generation agents, transmission agents* and *consumption agents*. First, generation agents broadcast their pricing information to all accessible consumption agents. Consumption agents predict their power needs in the future and place orders to generation agents. Here is a qualitative description of the logic of operation and coordination of these agents.

1. Transmission agents examine the expected load on the transmission lines.
2. If the loads are under the limit, no action will be taken. Otherwise, request consumers to reduce their power consumption to protect transmission lines.
3. A generation agent receives requests from consumptions agents and determine if it can satisfy all consumers.
4. If yes, no further action will be taken.
5. Otherwise, reject consumers' orders by.
6. Increasing price or providing a portion of requested power. A consumption agent receives modifying demand requests from transmission agents or generations agents, it can reduce its consumption or try to place order to other available utilities and start next cycle of negotiations.

Protocol is a critical issue in a distributed system. Some ideas can be adopted from Internet if we view the generation agents as hosts, transmission agents as routers and consumption agents as clients.

5.3.4 *Routing*

Routing is about the flow of the message within a network, e.g., how does a generation agent broadcast its pricing information to its customers and how is the order from a customer being transferred to the generation agent. A similar hierarchic naming strategy can be developed.

5.3.5 *Communication*

If the message from every agent can be propagated without any constraint, the entire network will soon crash because of the huge communication burden. Some sort of firewall mechanism should be built up to localize most of the communication. Transmission agents can be assigned this responsibility.

As discussed in Section 5.1, a key to achieving effective pruning at the grid-edge is using machine learning and AI carried by agents, so that the autonomy of the grid is enhanced. An important observation of the grid seen in Figure 5.1, is that it has loop or circuit structures at the transmission-distribution level and radial at the grid edge. In the modeling and AI space, circuits are graph structures where if you were to start at a certain node in the graph, there are pathways that will bring you back to the starting node; these structures are mathematically defined as totally ordered sets. Grid-edge data and deep machine learning modeling, on the other hand, may possess learning directionality guided by entropy production. The concept of "order" is important in deep machine learning, particularly in areas where hierarchical relationships, structured data, or a need for interpretability are present. Unlike totally ordered sets, partially ordered sets, or POSETs, are a mathematical concept that offers powerful tools for structuring data, designing more efficient and interpretable deep learning models, and providing theoretical foundations for machine learning problems at the grid-edge.

A Partially Ordered Set (POSET) is a fundamental concept in mathematics, pertaining to a of a set of elements and a binary relation (often denoted by $\leq$) that satisfies the following three properties:

1. *Reflexivity*: For any element a in the set, $a \leq a$. (Every element is related to itself.)
2. *Antisymmetry*: If $a \leq b$ and $b \leq a$, then $a = b$. (If two distinct elements are related in both directions, they must be the same element.)
3. *Transitivity*: If $a \leq b$ and $b \leq c$, then $a \leq c$. (If a is related to b, and b is related to c, then a is related to c.)

The key difference from a *total order* (like numbers on a number line) is that in a POSET, not every pair of elements needs to be comparable. For example, consider the set of subsets of a given set, ordered by inclusion ($\subseteq$). The set $\{1, 2\}$ and $\{2, 3\}$ are incomparable, as neither is a subset of the other.

The POSET ability to capture partial order relationships makes them particularly valuable in dealing with complex, hierarchical, and less-than-perfectly-ordered information prevalent in grid-edge machine learning applications.

Example 5.4. Graphs, POSETS and Congruence.

While in the physical grid, loops abound, in the AI space pertaining to the grid-edge, deep machine learning has directions, and

POSETs are an appropriate mathematical abstraction. Data-driven AI models, ordered in time as well as along other variables, e.g., entropy production, call for directionality that is given by POSETs. To illustrate, graph structures of complexity and the directionality of POSETs consider the patches and LAGs in Figure 5.3.

(a) Draw a graph to represent the LAGs in Figure 5.2(b).
(b) How many links are there in the graph?
(c) What is a matrix representation of the graph?
(d) What is a POSET on the graph?
(e) What is a congruence measure?
(f) Using Eqs. (5.1) and (5.2), calculate congruence in a garden under the (a) the usual measure of congruence given by Eq. (5.1), and (b) *Jaccard index* as given by Eq. (5.2).

Solution: Let us consider the 4 patches in Figure 5.3. In addition, their environment can be thought as another patch. Hence, we have a set of five patches $P = \{P_1, P_2, P_3, P_4, P_5\}$ which can be represented by the patches graph as follows.

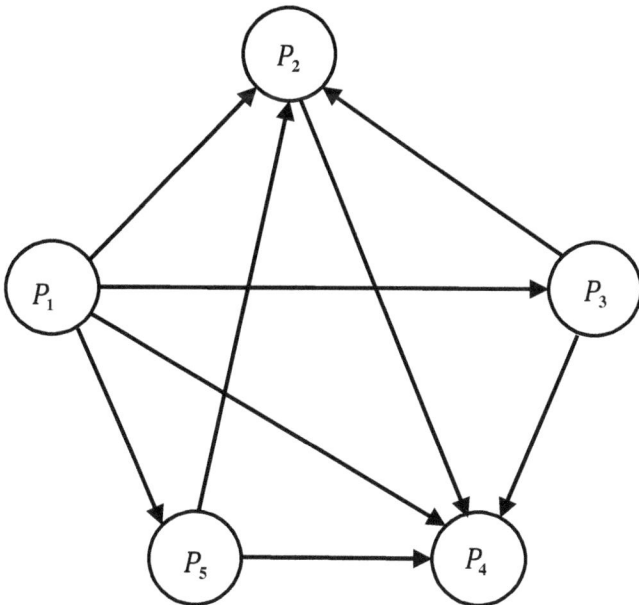

The matrix M representation of the patches graph is given by

$$M = \begin{bmatrix} 0 & 1 & 1 & 1 & 1 \\ 0 & 0 & 0 & 1 & 0 \\ 0 & 1 & 0 & 1 & 0 \\ 0 & 0 & 1 & 0 & 0 \\ 0 & 1 & 0 & 1 & 0 \end{bmatrix}$$

(a) We create here a graph representation of LAGs (Figure 5.3). Using the gardening metaphor, we can label LAGs as patches with the following labels:

5.3.6 *Nodes*

E: Environment

D: Daisy patch

S: Sunflower patch

R: Rose shrub patch

C: Tree canopy patch

5.3.7 *Directed edges*

$E \rightarrow D$

$E \rightarrow S$

$E \rightarrow R$

$E \rightarrow C$

$D \rightarrow S$

$D \rightarrow R$

$S \rightarrow C$

$R \rightarrow C$

(b) There are eight directed edges in total.

(c) To find the matrix representation we order the rows and columns as [E, D, S, R, C]. An assigned value of "1" means a directed link from the row node to the column node.

	E	D	S	R	C
E	0	1	1	1	1
D	0	0	1	1	0
S	0	0	0	0	1
R	0	0	0	0	1
C	0	0	0	0	0

(d) The underlying POSET structure is

Set representation: $P = \{E, D, S, R, C\}$

Relation R (representing the partial order "$\leq$"):

$$R = \{(E, E), (D, D), (S, S), (R, R), (C, C), (E, D), (E, S), (E, R),$$
$$(E, C), (D, S), (D, R), (S, C), (R, C)\}$$

The cover (Hasse) relations (omit reflexive and transitive "long" edges):

$$E \to D$$
$$D \to S$$
$$D \to R$$
$$S \to C$$
$$R \to C$$

The Hasse diagram (bottom up) is shown as follows:

```
      C
     / \
    S   R
     \ /
      D
      |
      E
```

(e) Let E_{graph} the set of the 8 edges in the adjacency graph and E_{POSET} the set of the five covering edges in the Hasse diagram. Thus, the following calculations can be made.

The usual measure of congruence is given by Eq. (5.1) is

$$C = \frac{|E_{\text{graph}} \cap E_{\text{POSET}}|}{(|E_{\text{graph}}| + |E_{\text{POSET}}|)/2}$$

If the Jaccard index is used, we have

$$J = \frac{|E_{\text{graph}} \cap E_{\text{POSET}}|}{|E_{\text{graph}} \cup E_{\text{POSET}}|}$$

(f) Based on our earlier findings, we have the following.

$$|E_{\text{graph}}| = 8$$
$$|E_{\text{POSET}}| = 5$$
$$|E_{\text{graph}} \cap E_{\text{POSET}}| = 5$$

Therefore, we can now calculate the numerical congruence under the two measures. The measure of congruence given by Eq. (5.1) is

$$C = \frac{5}{(5+8)/2} = \frac{5}{6.5} \approx 0.769$$

While the measure of congruence given by Eq. (5.2) is

$$J = \frac{5}{8+5-5} = \frac{5}{8} = 0.625$$

We observe that the two indices give comparable, but not the same, results. The Jaccard Index ranges from 0 (no common elements) to 1 (identical sets). Hence, the closer that J approaches 1, the more patches are similar in terms of their properties. If we are modeling LAGs as patches, when $J = 1$, it is possible to integrate several of them in a new "equivalent LAG." This is something analogous (in the grid-edge agent space) to what electrical engineers do in modeling through equivalent circuits such as *Thevenin and Norton equivalents* which under certain mathematical conditions can adequately capture the aggregate behavior of electric currents and voltages in AC circuits.

Example 5.5. POSETs and Simple Grid Dependence.

A POSET consists of a set of elements equipped with a binary relation ($\leq$) that satisfies *reflexivity, antisymmetry,* and *transitivity.* Unlike a total order, such as numbers on a line where every element can be compared elements in a POSET need not be universally comparable. This characteristic allows POSETS to represent systems with hierarchical structures or directional dependencies effectively. In the context of energy transitions, especially when employing machine learning and transfer learning, POSETS help visualize and manage complex dependencies in the modeling workflow.

Consider a simplified scenario involving machine learning for grid-edge applications and show how POSETS may be used if the workflow involves three distinct modeling tasks:

A: Initial Data Acquisition – collecting and preprocessing raw data from various sensors.

B: Feature Extraction – transforming raw data into meaningful features.

C: Predictive Modeling – using extracted features to predict grid behavior.

Solution: The dependencies between the tasks reflect a clear hierarchical order. Task B cannot proceed without completion of Task A, while Task C cannot proceed without completion of Task B. We have:

$A \leq B$: Feature extraction depends on data acquisition.

$B \leq C$: Predictive modeling depends on feature extraction.

This scenario forms a clear POSET structure. Consider,

Set representation:

$$P = \{A, B, C\}$$

Relation R (representing the partial order "$\leq$"):

$$R \leq= \{(A, A), (B, B), (C, C), (A, B), (B, C), (A, C)\}$$

5.3.8 *Matrix representation*

	A	B	C
A	1	1	1
B	0	1	1
C	0	0	1

The Hasse diagram visually represents these dependencies without showing the implied connections from reflexivity and transitivity, simplifying understanding of direct task relationships. The Hasse diagram in this example is shown as follows:

$$
\begin{array}{c}
A \\
\downarrow \\
B \\
\downarrow \\
C
\end{array}
$$

Example 5.6. POSETS and Transfer Learning Hierarchy.

In machine learning applications for power grids, transfer learning can leverage knowledge from generalized models to improve specialized scenarios. Consider three machine learning models below and show they form a POSET.

X: General baseline model – trained on diverse datasets covering multiple grid conditions.
Y: Urban-specific model – trained specifically for urban power grid environments.
Z: Rural-specific model – trained specifically for rural power grid environments.

The transfer learning relationships are Model Y uses insights gained from Model X, and Model Z uses insights gained from Model X.

$X \leq Y$: Urban model builds upon baseline knowledge.
$X \leq Z$: Rural model builds upon baseline knowledge.

Solution: Here, Model Y and Model Z are independent specializations; neither directly influences the other, making them incomparable. Therefore, this POSET can be represented as

Set representation:

$$P = \{X, Y, Z\}$$

Relation R (representing the partial order "$\leq$"):

$$R = \{(X, X), (Y, Y), (Z, Z), (X, Y), (X, Z)\}$$

5.3.9 *Matrix representation*

	X	Y	Z
X	1	1	1
Y	0	1	0
Z	0	0	1

The Hasse diagram clearly illustrates the foundational role of the generalized model (X) in supporting specialized adaptations for distinct environments (urban and rural), effectively representing the concept of transfer learning.

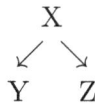

$$
\begin{array}{ccc}
 & X & \\
\swarrow & & \searrow \\
Y & & Z
\end{array}
$$

These examples illustrate the powerful applicability of POSETS in clearly expressing dependencies and directionalities common in energy transition machine learning scenarios. While actual grid infrastructure inherently includes loops or circuits, machine learning pipelines leveraging data from such grids typically exhibit clear hierarchical or directional structures. Thus, POSETS offer a valuable mathematical framework for representing and managing these relationships, enhancing clarity, robustness, and efficiency in machine learning model development and deployment.

5.4　From Smart Devices and Processes to Smart Energy

Despite the ethereal connotation of the cloud, it is a major hardware infrastructure with a very large number of computers, concentrated in remote areas or distributed in near-the-users data centers running AI applications. Same is true for data centers. The electricity requirements to run them are extraordinary. It is no hyperbole to say that AI represents the most energy intensive application of computing.

On the other hand, ever since the revolution that information technology has brought about, smart solutions have emerged as a class of techniques that utilize better sensing, advanced information processing and intelligent decision-making to tackle a plethora of challenging problems. Smart solutions in energy systems are a necessity.

Smart appliances, smart energy management tools, smart meters, and smart grids are labeled "smart" to highlight some of their innovative features. For example, smart appliances, such as washers/dryers, differ from the "dumb" versions in that they may be informed in real time about the price of the electricity they use, or are about to use, and therefore may be able to adjust their operating cycles to lower the overall cost of energy. When we have many smart appliances, for example inside an office building, it is convenient to have a unified interface to connect and manage all devices, a tool generally referred to as smart energy management. For smart appliance/smart energy management to receive real-time price information, smart meters are introduced in the grid-edge. Smart meters are devices equipped with two-way communication capabilities to allow utilities to send out real-time pricing signals to users and to monitor electricity usages remotely. Ultimately, all these smart solutions combined with better sensing and control on the transmission and distribution side, participate as elements of a "smart grid."

Smart processes are implemented through smart devices that abound in a wide range of technical and human activities including telecommunications and computer networks, banking and finance, environmental monitoring, industrial processes and of course power. The popularity of smart devices has adequately demonstrated a well-accepted vision that devices with sufficient degree of "smartness" may be silver bullets for hard problems. This expectation is valid mainly because smartness is vaguely defined, and many instruments can be labeled as "smart devices." So, the question is what devices can and should be labeled as smart? This question can be answered in two different ways.

In a popular and nontechnical way, smart can simply mean "advanced" or "better." For example, "smart" phones are widely used by manufacturers and wireless providers to differentiate their product lines from the old ones. Smart phones are anything but phones. They are hand-held personal computers with myriads of functionalities embedded in a device that is called "phone" mostly for marketing purposes – to create a sense of familiarity and continuity with old telephones.

Technically, smart devices are systems that employ AI technologies (including relevant guidelines) for data-driven inferencing to assist problem solving. In this definition, intelligent methodologies are referred to as algorithms that can reprogram or reconfigure themselves to achieve a set of desirable goals with minimal human interventions. Simply put, they are devices that can operate *autonomously* in the foreground and undergo machine learning and evolution in the background. Both machine learning and evolution are attributes exclusively characterizing intellectual entities, especially human beings. If a device has such capabilities, it can be called smart.

To explore the AI-energy nexus, we have collected a list of applications of smart solutions in the grid-edge.

5.4.1 *Smart appliances*

In the old days most of us were paying a flat rate for electricity. When and how to use our appliance were never a serious issue. A monthly review of the electricity bills was sufficient for a good estimate. However, in electricity economics the flat-rate model does not provide a good pricing structure because it does not reflect the true cost that is determined by factors as demand, supply and delivery capacity, which fluctuate from minute to minute. It is widely accepted that this disconnection of prices from costs is not healthy and needs to be corrected with a real-time dynamic pricing model that reflects the true cost. With the real-time dynamic pricing model, electricity price can vary by a factor of ten depending on the time of the day.

It is unrealistic to expect customers to monitor prices around the clock. Therefore, appliances in the future need to be operated smartly and to make the right adjustments automatically according to the actual pricing information. For example, scheduling some noncritical activities, such as laundry, to off-peak hours when the price is low is to good way to reduce the electricity bill.

5.4.2 *Smart energy management tools*

Smart appliances can operate automatically, making decisions on behalf of customers. However, they cannot be autonomous; they require inputs from users who need to monitor their performance as well. As the number of devices grows, for example in an office building, a smart energy management system is a convenient tool for users to interact with devices. The smart energy management system connects and controls all smart appliances; it also provides a unified interface for programming users' preferences straightforwardly and naturally. For example, a user will prefer to program an AC unit to "turn off when the price is high." This natural language will have to be translated into a numeric representation (e.g., dollar per kwh) for a machine to understand. Some smart energy management tools also include features such as a web portal, from where users can quickly access their energy profiles.

5.4.3 *Smart meters*

For smart appliances to work, they need to receive real time pricing information from the suppliers. Older versions of electric/gas meters can only transmit data back to the utilities. Smart meters, on the other hand, have a two-way communication capability, which allows utilities to relay real-time pricing data to the users.

5.4.4 *Smart grids*

Ultimately, the true benefit of the smart solution can only be revealed when all smart devices are connected within a network called smart grid. The smart grid is a collective term covering innovations on two layers, physical (or hardware) and information (McDonald and

Thomas, 2019). In the physical layer, a *smart grid*, aside from integrating smart meters and appliances, implements enhanced reliability and safety features to the transmission and distribution network, such as self-diagnosis and self-healing capabilities via advanced electronics and signal processing. In the information layer, a smart grid will operate under new business/market models, such as the real-time dynamic pricing model. Another marketing scenario a future smart grid has to include is role change between buyers and sellers. A customer driving a hybrid, or an electric vehicle is a buyer when he charges his vehicle. He can become a seller when he decides that the price is high for him to make some profits by selling the extra capacity back to the grid. Similarly, a family with a wind turbine is a net consumer in the windless days; it can become a net provider when wind is abundant. A sophisticated market model is needed for incorporating all these variables.

The short list above demonstrates how versatile smart technologies are. It also reveals the fact that the "smart" label is used almost indiscriminately. This convenience becomes confusing and problematic when it comes to system integration of smart grids that involve smart devices. Until there is a consensus among the *device world* on what is "smartness" the smart grid will be no more than a collection of individual components. To illustrate the point, if a smart device A is in the same network with another smart device B; unless both devices speak a common language, they will not be able to work cooperatively (McDonald and Thomas, 2019).

Defining a formal language for smart devices to "talk" requires a vigorous standardization process, which is beyond the scope of this book. Instead, we focus on discussing the core issue: Define smartness in a logical and implementable fashion. These discussions intend to be conceptual and inspirational, rather than formal and comprehensive. The goal is to develop a prototype metric for measuring smartness quantitatively and generically. A quantitative metric uses numeric scales to characterize the degree of smartness, like the IQ test. This feature is the cornerstone for the evolution of smart devices. They can become smarter only if their smartness can be measured qualitatively. A generic metric is applicable for most devices. This feature is a necessity for system integration. With this feature, smart devices A and B are comparable and comprehensible.

Smartness, also referred to as intelligence, is defined as the capability of making humanlike adjustments in response to changing environments. AI brings the science and technology of understanding human intelligence to bear on building intelligent entities to perform purposeful tasks. In the modern computer age, the intelligent entity concerned is usually implemented with a piece of software program, also called intelligent agent. For example, a smart appliance is controlled by an intelligent agent, a computer program running on a computational resource inside the appliance. The intelligent agent receives pricing signals from a smart meter, analyzing the information, and controlling the appliance's operating cycles. This example summarizes three basic functionalities of an intelligent agent: sensing, decision making and acting. Researchers from the AI area have summarized four possible ways to identify a computer agent as being intelligent: acting humanly, thinking humanly, acting rationally, or thinking rationally. A common attribute shared by these four criteria is the ability to generalize.

An agent with generalization capability learns from known examples and creates new relations (knowledge) in ways that unknown inputs can be handled appropriately. The smart appliance, for example, may have learnt to turn off when electricity price is high and turn on when price is low. This knowledge (rule) can be programmed into the intelligent agent that controls the appliance, a process we call expert learning. After learning, the agent will know how to operate when the input is one of the two known ones: "price is low" or "price is high." A "smarter" agent, on the other hand, can respond to novel situations such as "price is medium." To do that, it has to interpolate from relations it has already learned such as, "price low" to "appliance on" and "price high" to "appliance off," and make inferences on a new relation, something like "price medium" to "appliance operating on 50% cycle." Achieving this level of smartness requires a tremendous amount of effort for a computer.

How to determine whether a device is smart or not and what is the degree of smartness? To answer this question we need a measurable definition for smartness. From a technical perspective, measuring smartness is as important as defining it. We propose an empirical definition that leads to measurement.

5.4.5 *Capacity-based measures*

One way to evaluate the smartness of a device is via its design. Using this criterion, a modern computer system, consisting of microprocessor, memory, I/O components, is significantly more capable (and thus smarter) than a simple circuit breaker, which includes only fuse, resistors and capacitors.

The capacity-based metric, determined by a device's projected achievement rather than its actual performance, has some significant advantages, such as early applicability and situation independence. Since it's measured from the design, once the implementation details of a device are ready, its degree of smartness can be evaluated immediately. This is a big benefit for system designers who often must work on components that are still in development stages. Also, this metric is not affected by the prospective applications of the device. Once recognized as a smart one after a thorough examination and verification process, a device is considered smart for the rest of its life no matter what environment it works in.

The disadvantages of the capacity-based definition are also evident. First, it is a subjective rather than an objective metric. The superiority of a design is often conditional on the assumptions and is not directly comparable to other options. Second, it is hard to quantify. Even if two devices are comparable under the same assumptions, the result would be categorical rather than numerical. It is easier to determine A is smarter than B, but a lot harder to conclude by how much. Third, the evaluation process is long and resource exhaustive. The examination and verification process will be a heuristic one, requiring the presence of human experts or qualified parties (other smart devices. AI works well with natural intelligence, to make experts more efficient in the use of their expertise, and novices become experts faster and better.

5.4.6 *Performance-based measures*

An alternative method to investigate intelligence is through observing the interactions of an entity with its environment. As advocated by researchers as AI, intelligence is revealed at the human-machine relation (Tsoukalas, 2023). This concept agrees with Alan Turing's original idea of using a set of carefully organized questions to determine whether a machine is intelligent or not. In the Turing Test, if a machine can answer questions in a way that an interrogator cannot tell if it is machine or human, this machine is intelligent. Expanding this idea to determine how smart a device is, we can look at its performance as it responds to a set of inputs or questions. A device's degree of smartness can be graded based on the results. The higher the score, the smarter the device.

The performance-based metric is thought to be more accurate and objective than the capacity-based metric since questions can be designed to reflect the actual application scenarios. It is also independent of application details, which lack uniform standards across different devices. The best feature the performance-based metric offers to engineers is the numeric output. With that we can compare two devices quantitatively.

As in the Turing Test, a potential problem is how to develop the questionnaire accurately, adequately, and objectively. The question development process requires substantive and full knowledge of the application scenarios, which makes it more appropriate for an *a posteriori* evaluation. This presents a significant challenge for system engineers who often need an *a priori* metric to apply during the design phase.

5.4.7   *Complexity-based measures*

Motivated by the preceding discussions, we propose a metric that is as predictive and generic as the capacity-based measure and as objective and quantifiable as the performance-based measure. To achieve a workable balance of these goals we propose switching our focus from "what intelligence is" to "what intelligence is for." The answer then becomes, that in a complex system, where the interactions among components are more important than the number of components or parts, intelligence is used for handling complexity. In physics, the three-body-problem is the largest one we can handle analytically. Even such a "small" problem can reveal so much complexity that we still do not understand it fully. For a system such as an electric power grid, which involves thousands of generators, millions of miles of lines, and millions of customers, even numerical simulations are nearly impossible without considerable simplifications. The introduction of intelligence can aid handling interactions leading to greater system manageability. To this regard, studying smartness and smart devices in an isolated manner without addressing the interactions with their environment is meaningless. Therefore, it is logical to define intelligence of a device as its capacity of handling interactions.

The most economical way to evaluate the capacity of a device to handle interactions is by examining its input vector. The higher number of dimensions the input vector has, the more capable the device is. Potentially, *dimensionality* can be used as a numerical measure of smartness.

The intelligence of a device or a system can be evaluated based on its ability of managing complexity. This ability is equivalent to its capacity of handling interactions with its environment. This can be further characterized, in a simplified and generic manner, from the device's inputs. Qualitatively, the more information a device extracts and utilizes from inputs, the better capacity it must manage complexity. Numerically, one of the simplest ways to define the degree of smartness is the number of dimensions on the inputs.

An important remark on this definition of intelligence is that both capacity and interactions were used in the same sentence. This is also a hint that the complexity-based metric is a hybrid of capability based and performance-based metrics, which allows us to take advantages of both definitions.

First, defining a device's smartness based on its capacity rather than entirely on its performance is necessary based on two considerations. One, capacity is a more accessible and generic measure than performance. Two, performance is case (inputs) dependent and does not fully capture the whole dynamics of the device. Performance can be improved over time through learning, which is another characteristic of smart devices.

Second, defining smartness of a device based on its input dimensionality is a crucial step to include the interaction element. This brings two additional benefits. One, it is very close to being implementation independent, with excellent comparability and compatibility among devices with different designs and algorithms. Two, it does not derive from the performance of the device, which extends its applicability universally to devices or systems even in a design phase.

Using this interpretation, we now have a comparable metric to be applied universally over all smart devices. For example, a device that takes voltage and current as its inputs will be smarter if it also considers a third dimension, such as frequency. Likewise, a smart meter is smarter than a conventional one, in that it uses time as another dimension to differentiate electricity prices. Perhaps this is the most valuable message: To be intelligent, one (human or device) must be able to differentiate inputs because all other elements (processing, decision making, etc.) will be based upon several recognized patterns. Therefore, it is a natural conclusion that, an energy system could not be smart until devices can differentiate the object they handle, energy (or electricity). In current models, electricity is defined as a physical quantity without identity, which is what we try to improve in the next section.

As the power grid acquires significant learning and self-healing capabilities to adapt and reorganize itself in evolving conditions, we can work with pockets of electric energy and data, bundled together in an innovative structure known as *smart energy*. There are compelling needs for such innovations especially at the grid-edge due to major shifts towards electricity. There are several reasons for that. First, the massive introduction of distributed energy resources like solar, wind, biomass, geothermal and grid batteries. Second, environmental concerns about climate change may include in energy prices externalities, such as emissions costs and penalties giving rise to new business models for energy. To absorb these and other changes, our energy systems will get more electrified, that is, electricity transmission and distribution networks face growing demand and call for innovations. The addition of intermittent renewable sources significantly increases the burden on energy systems to handle extra stochasticity on the generation side. Any stochasticity, if not managed appropriately, could lead to reliability and safety issues. Smart solutions can address these concerns in autonomous or semi-autonomous ways, and in doing so improve reliability, efficiency, and safety. It should be noted however that from a customer perspective, low cost, and minimal or no deterioration in energy quality and availability are prioritized to environmental concerns. From the suppliers and marketers' perspectives, the solutions should align with policy and regulation while creating acceptable returns on investment.

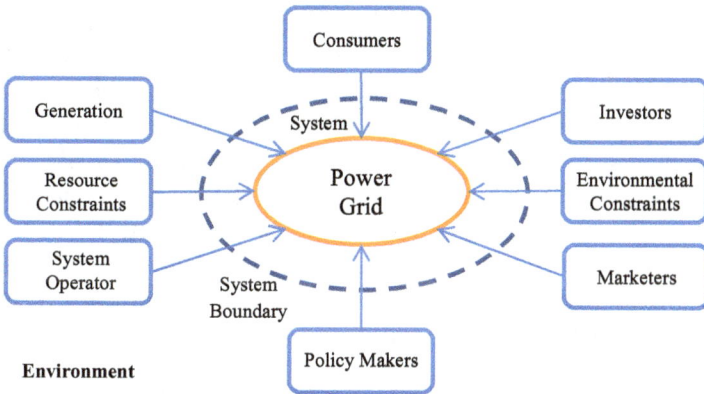

Figure 5.11. Power grid centered modeling with procedure-oriented programming.

To accomplish this goal, conventional power-grid modeling approaches need major modifications. In such approaches, the power grid sits at the center, blanketed with systems to be modeled is shown in Figure 5.11. Most relevant variables interface with the grid as environment variables. A boundary is drawn between the system and its players, such as generation, consumers, investors, system operators and policy makers. The arrangement shown in Figure 5.11 is efficient for a conventional process-oriented (or procedure-oriented) (POM) modeling perspective using a top-down approach. Here, major components are identified first and then their functional descriptions developed before the design is refined to the implementation level, where data structures can be defined. The main advantage of POM is its efficiency. With succinct structural organization and functional definition, it is suitable for smaller systems, where central planning philosophy is applicable. It is not a coincidence that POM has been the primary method used in energy systems in the past because the old procedure was simpler: price was flat; operations were highly regulated, and supplies were predictable.

However, when a system expands in size, many devices operate and interact simultaneously therefore the order of possible scenarios grows exponentially. In these cases, it is difficult to clearly define device functionality at a system level. We need a bottom-up approach, pivoting on the interactions rather than the components. This approach is called object-oriented modeling (OOM). In contrast to POM, OOM starts from defining objects that link all components together. In OOM, an object is an entity with "states" and "behaviors," corresponding to "properties" and "methods" respectively in programming languages. For example, in a human resource database, "employee" can be defined as an object. Name, Date of Birth, Social Security Number, Salary, etc., are all candidate properties describing the employee. Likewise, Creation, Promotion and Deletion are possible methods that affect one or more properties of an employee.

Viewing an object as an abstract encapsulation of both properties and methods offers several advantages. One, it provides modularity. Different types of objects can be defined separately and independently. Two, it masks unnecessary information. Since both properties and methods are defined inside the object, unnecessary details, e.g., properties and methods for internal use only, can be hidden

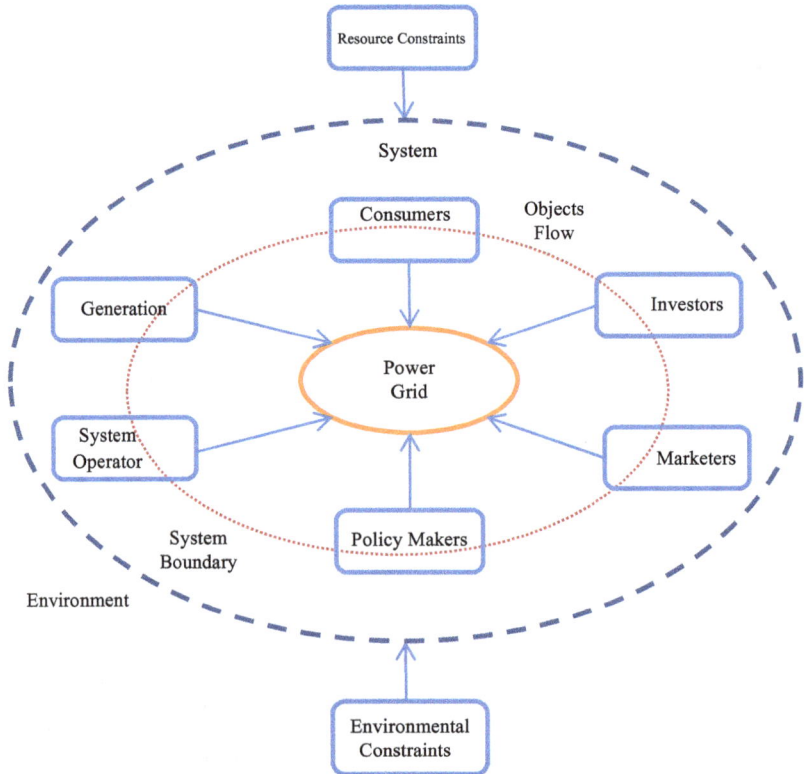

Figure 5.12. Smart-energy modeling with object-oriented programming.

from outside access. Both benefits are particularly valuable for large system modeling, such as an energy system. In this respect, OOM is superior to POM because the dynamics of the system can be naturally captured through the process of object construction, instantiation, modification, and deletion.

A smart energy system using object-oriented modeling is shown in Figure 5.12. With OOM instead of POM, we can expand the system boundary to include important components such as investors, generation, system operators, consumers, and policy makers. To use object-oriented modeling techniques, we need to define objects that balances demand to available energy. Smart energy differs from the "not-so" smart one in the following:

(1) More properties and attributes are defined.
(2) Methods use advanced algorithms to utilize the additional dimensions to program more activities such as the intention of the object.

There is more than one way to define smart energy but from the system perspective it should at least include the items shown in Table 5.2. Most of the attributes and methods are self-explanatory for ordinary objects. However, for electricity, its temporal dimension

Table 5.2. Sample data structure for smart energy objects.

Smart Energy	
Properties	
Amount	*Amount of t physical energy in kWh*
Price	*Selling price in dollars*
Producer	*Utility ID*
Route	*Transmission and distribution route*
Seller	*Seller ID*
Buyer	*Customer ID*
Current time	*Current time stamp*
Generation time	*Scheduled production time*
Consuming time	*Scheduled consumption time*
Environmental cost	*Added environmental impact*
Methods	
Construction	*Method to initialize an object*
Destruction	*Method to destroy an object*
Modify	*Method to modify an object's properties*

must be considered with extra caution taken because electrons travel almost at the speed of light, which leaves very little room for cushion. This is problematic for implementing smart energy networks both methodologically and technically.

5.5   Anticipation to Reduce Entropy Production

In grid-edge, the anticipation of future states can significantly contribute to self-healing approaches and in a more general sense reduce entropy production. This is because proactive actions, informed by trustworthy predictions, can prevent the development of highly entropic states, or steer the system towards more ordered and less dissipative pathways. Analyzing possibilities for how this would work we can draw parallels with the pruning metaphor. For grid-edge optimization, anticipating future states is crucial for efficient entropy management. Consider the following.

Predicting renewable energy generation: Knowing when solar or wind generation will be high or low allows for proactive management of energy storage and demand response, minimizing energy waste and grid instability.

Forecasting demand fluctuations: Anticipating peaks in energy consumption enables the deployment of resources and control strategies to meet demand efficiently, avoiding sudden surges and potential blackouts (high entropy events).

Predicting EV charging patterns: Understanding when and where EVs will charge allows for optimized charging schedules that benefit both EV owners and the grid, preventing grid congestion and inefficient energy use.

Anticipating potential equipment failures: Predictive maintenance on grid-edge devices (inverters, batteries, transformers) can prevent failures that would lead to energy losses and increased entropy.

In general, anticipation can prevent overgrowth and inefficiencies. Without anticipation, a gardener is acting in a reacting mode, for instance, prunes only after a plant becomes overgrown and tangled. Similarly for the grid-edge. If managed without anticipating future states often deals with inefficiencies and increasing disorder reactively. This reactive approach can be energy-intensive and lead to higher entropy production in the process of correction. On the other hand, with anticipation, that is when grid-edge agents can reliably forecast potential future states (e.g., predicting high energy demand, potential system failures, or resource depletion), preemptive actions can be taken to avoid these less desirable, high-entropy states. This is akin to a gardener proactively pruning young shoots to prevent overcrowding later, which is less disruptive and resource-intensive than dealing with a heavily overgrown plant.

Optimizing resource allocation over time, if done without anticipation, resources might be allocated based on immediate needs, leading to suboptimal distribution and potential waste in the long run. This can result in higher entropy production due to inefficient use and dissipation of energy and materials. On the other hand, predicting future resource demands allows for optimized allocation over time. For example, anticipating a surge in energy demand can trigger the pre-positioning of energy reserves or the activation of more efficient generation sources.

Steering the system towards lower entropy production trajectories, on the one hand, without anticipation, the system evolves passively, potentially drifting towards states with higher disorder and entropy production due to internal dynamics or external disturbances. On the other hand, with anticipation, we can implement control strategies that actively steer the system along trajectories characterized by lower entropy production. This is like a gardener training a plant to grow in a specific, organized way, preventing chaotic growth patterns.

Several additional areas of benefit to the grid-edge from anticipation can be identified. For instance, when it comes to implementing predictive maintenance and failure prevention without anticipation failures are dealt with reactively, often involving significant energy expenditure for repairs and replacements, and leading to increased entropy due to system downtime and material waste. Predicting potential failures through monitoring and modeling allows for proactive maintenance. Addressing issues before they escalate prevents catastrophic failures and the associated high entropy production from system breakdown and emergency responses. Furthermore, systems designed without considering future uncertainties can be brittle and prone to collapse or inefficiency under unforeseen conditions, leading to sharp increases in entropy production during failure and recovery. Incorporating predictions of potential future scenarios (e.g., climate change impacts, market fluctuations) into system design allows for the development of more resilient and adaptive systems. These systems can proactively adjust to changing conditions, maintaining lower entropy production over the long term.

In conclusion, anticipating future system states enables proactive and informed decision-making, leading to more efficient resource allocation, smoother operation, and the prevention of high-entropy events. By "pruning" potential inefficiencies and steering the system towards more organized and predictable pathways, we can effectively manage and potentially reduce the overall rate of entropy production associated with the system's function. This proactive approach is fundamentally more efficient than reacting to problems after they have already manifested.

On the generation side, natural gas generators have a prominent role as they respond quickly to wind and solar generation variabilities. Natural gas turbines run on a Brayton cycle which has greater flexibility and speed in following demand fluctuations. Biomass and geothermal energy can also contribute flexibility. Yet, there is little doubt that atomic power will be needed, at scale, for a sustainable path forward. Accelerating the Atomic energy transition through innovations in advanced technologies such as small modular reactors (SMRs) and microreactors (MRs) as well as the construction of new full-size light water reactors (LWRs) is *sine qua non* for viable energy strategies under AI. But this is a somewhat long-term option, it will take at least a decade. It does present though new and unique challenges. For instance, SMRs (up to 300 MWe power output) and MRs (up to 10 MWe power output) may be used to provide power exclusively to data centers, or, steam-to-process industries. In such energy ecosystems (microgrids), load following would be an important requirement, and new reactors would need to be designed to go well beyond the 5% load following capacity that present reactor designs have. In addition to such operational constraints addressed in the design phase, the new technology can utilize AI to develop semi-autonomous anticipatory control approaches to meet performance requirements with the least possible aging and degradation effects in the lifetime of the reactors.

In operational terms, the effective coupling of the power grid with AI rests on a simple but potentially powerful idea: *To protect itself from upset events, the grid should act proactively, that is, affect local control in anticipation (not just in response) of possible contingencies.*

The thesis that anticipations of the future are an integral part of control strategies originated in the field of mathematical biology and ethology by the pre-eminent biologist Robert Rosen as a way of explaining how organisms cope with complexity (Rosen, 1985). The anticipatory approach will allow the grid to protect itself from local faults (that may be likely to cascade to future global failures) through the combined use of AI predictive models and fuzzy control strategies emulating the synergism of perceptual and linguistic computing performed in the human brain (Tsoukalas and Uhrig, 1997).

In the short term, to meet AI demand while ensuring grid stability, the AI-Energy nexus may provide solutions for the efficient use of energy. For instance, with a stronger coupling of power with AI, *smart energy* improves the integration of diverse energy resources and paves the way for a resilient, affordable, and sustainable energy infrastructure undergirded by emissions-free, dispatchable atomic energy. As an energy strategy to leverage natural gas (short to medium term), embracing smart energy technologies (short term), paves the way for the Atomic energy transition (longer term).

Grid-edge and smart energy are not physics terms like entropy, enthalpy, energy, and the like. They are concepts linking energy with AI in a multifaceted way adding value to inferencing support for energy. For instance, for the grid of the future, we need to forecast energy demand in a multitude of horizons that range from seconds to minutes and hours, but also days and years into the future. Forecasting tools use large volumes of data from demographics, economics, fuel prices, capital costs, licensing, and technology options which with the help of AI become easier to modify and transfer to future planners.

Something analogous happens at the other end of the forecasting scale. For instance, to forecast weather in short-time and very short-term future horizons. Here, AI can be a protagonist. Data-driven models can learn about local idiosyncratic patterns of supply and demand and can address the risks of weather variability; cataclysmic events such as floods or wildfires or previously unseen anomalies in weather conditions. At the grid-edge, myriads of decisions affecting plans, schedules and costs, prices and gains, can be integrated with energy and information in real-time and in transparent and easily modifiable ways.

Smart energy rests primarily on AI's capacity to anticipate. In Figure 5.1, electricity travels from generators to consumers at the speed of light. Hence, predicting or anticipating the future, in effect increases the time to schedule based on scenarios that are likely, and meet anticipated conditions successfully. In a more fundamental sense, AI is a knowledge network running on scenarios produced by memory and communication capacities. Hence AI is fundamentally limited by bandwidth allocation. On the other hand, electricity flows in power systems limited by peak demand. A power system must meet peak-demand no matter and hence may end up wasting energy in anticipation of a never certain peak demand.

Hence, AI and power are two different types of networks, the first bandwidth constrained, the second peak-demand constrained. They can be harmonized through AI's capacity for anticipation. For instance, through AI's capacity to anticipate, the AI-Energy nexus can address the difficulties of energy storage (a hard physical problem) by transforming storage into a problem of scheduling which can be solved though computational approaches (Tsoukalas, 2023).

In a technical sense, smart energy can be thought of as a computer object, a vector whose attributes include, but are not limited to, energy and fuel, price, time of production, time of consumption, exogenous costs, emissions, reliability, pollution costs, regulatory information, as well as market and environmental constraints. From generation to consumption, the flows of data about energy form an extremely complex system of upstream and downstream ventures, facilities, and civil, defense and industrial infrastructures including the AI infrastructure of data centers and the cloud. Along all these complex energy flows, decisions are made, either through engineering design and automation or through human intention and choice. Energy infrastructures, like their data and information counterparts, exhibit network behaviors where network functionality is affected by connectivity and choice.

An illustrative technical framework for smart energy network is outlined next. First, we define smartness as the ability of a device/system to manage complexity and quantify it using the dimensionality of the inputs. Second, smart energy is described as an object with attributes and properties such that object oriented modeling (OOM) techniques can be applied to complex energy systems. Next, using electricity as an example, we show that a requisite for modeling energy as an object is the storage capacity, a problem whose novel solution consists of a virtual buffer (essentially a metaphor for scheduling). Finally, we present illustrative examples of how smart energy works in practical terms.

Optimal generation has become very important with the introduction of the deregulated power markets. The previously vertically structured power industry is in the process of unbundling into three principal groups: generators, transmission owners, and independent system operators. In this new setup, all three parties need to operate at optimal or close to optimal level. Traditionally, advanced optimal operations were based on the Optimal Power Flow (OPF) introduced some forty years ago. Although the idea is simple to express mathematically, it is a very difficult optimization problem. The classical OPF formulation has been complicated even more with the new requirements of the deregulated market. In our research, we will address some of the new requirements and how they can be addressed by an OPF based on the Artificial Intelligence (AI) principles. The main goal is a working OPF that can be used for other optimization problems such as Unit Commitment (UC). Initially, operational feasibility was confirmed on a series of small networks. Results were compared with available commercial OPF packages using objective functions that could be handled by these packages. After it was obvious that GA can be used for OPF, the whole algorithm was rewritten to handle large, real-life systems. The objective function of interest is still to be determined from close interaction with power producers and transmission controllers.

Electric generation control is performed in a distributed manner to supply power to geographically defined control areas. The goal of generation control is to keep the inadvertent flow of power across a control area's boundary as small as possible. If a difference exists between the power supplied and the power demanded in a control area, the load deficit or surplus would be either borrowed from or stored as the kinetic energy in rotating machines in the grid.

An illustrative example can be seen in addressing the challenge of matching the power demand of a LAG with the power delivered by a coal-fired power plant. An anticipatory controller for a model power plant is presented to prescribe the power output into the grid. The control system forecasts what the future demand of the power customers in a control area is likely to be and modifies the fuel input to the power generation facility to match the predicted demand.

A neural network was found to be an adaptable and robust prediction mechanism for the highly nonlinear data found in the power consumption patterns of residential loads. The corresponding control schedule of the power plant was tuned to match the anticipated demand using an iterative neural network approach. The use of neural networks and an iterative scheme allows the controller designed in this research to be applied to a broad range of control problems. The control methodology considers limits in the magnitude and rate of control actions.

Simulations show that this implementation of anticipatory control of electric power demands is effective and especially well-suited for dynamic systems that include a dead time or control limitations. Such is the case for SMRs and microreactors but also for coal units of comparable size (e.g., 80 MW). The response of the anticipatory neural network control system was shown to be more energy efficient than feedback control for these generators and to have a much smoother, reduced control effort.

While anticipatory systems are and have always been an integral part of everyday life (and biological systems), the study of anticipatory systems is a relatively new field that comes about from the success of data-driven modeling. A simple example of an anticipatory system is a driver of an automobile slowing down in a residential neighborhood where children are playing. The act of reducing the car's velocity allows for a reduced stopping distance should one of the youths jump in front of the car. This happens virtually without thought from the automobile operator. By driving more cautiously the safety of the car/children system is improved and can then be transferred to an autonomous driverless car.

Can anticipations be exploited to make engineering systems perform better, or in a more safe, more flexible, and generally more optimal manner? To address this question in a technically sound fashion, a new system for predicting future states of a control system operating in an environment is proposed and examined through application to the deregulated electric power system and comparisons with conventional approaches.

To describe an anticipatory system in an abstract and more general manner, we view it as a system that utilizes predictions states of itself and of its environment to direct its present actions (Rosen, 1985). In the example, the driver of the car knows both the trajectory of his automobile and the possible trajectories of any children who are playing in the area. The act of slowing the car down is non-causal and is based on anticipated states of the entire system.

Two systems are needed to study the usefulness of anticipatory regulation applied to the electric grid. The first is the power consumption data for a given control area, and the second is the mathematical simulation of the power producing facility. Figure 5.13 shows the power demand recorded for a specific residential feeder. The feeder serves 2311 single-family dwellings and 1655 multi-family units such as apartment buildings. The power data is sampled every 15 minutes. Figure 5.14 shows a schematic of the generator as a general model (meaning the least number of empirical relationships used for calculation). The work is a component-based, first-principle

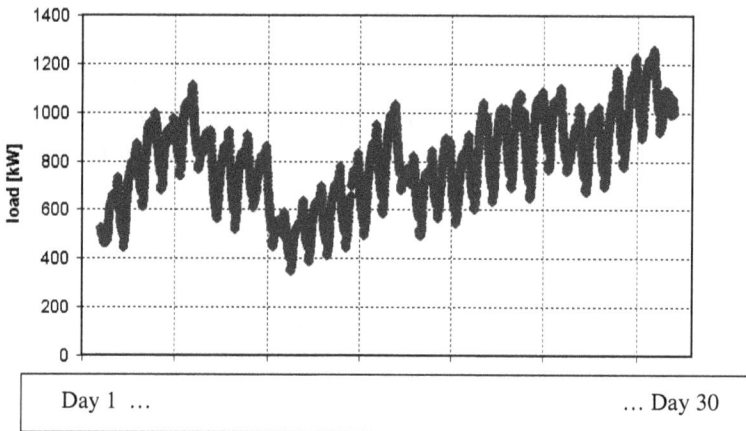

Figure 5.13. Power demand on a residential feeder line for 30 days.

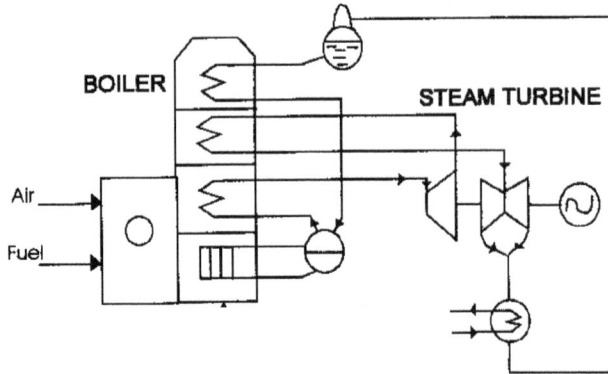

Figure 5.14.   Generator (boiler) schematic.

implementation of a pulverized coal electrical generator studied by Dr. Thomas Fieno.

The forecasting methodology is based on neural networks which have been proven to be superior nonlinear forecasting systems (Tsoukalas and Uhrig, 1997). In addition to this, it is a simple matter to incorporate adaptability and other variables of interest into the forecasting system due to the pattern matching functionality. Neural networks are chosen as the forecasting system for this study.

The network architecture was chosen as a fully connected, feedforward neural net with a "tansig" hidden layer activation function and a linear output layer activation function. Due to seasonal variations of electric power data, it is also advantageous to have an adaptability system in place. To accomplish this a concurrent neural network trained on more recent data than the primary network was implemented. When the second network consistently (for three days of data) predicted more accurately than the primary network, the weight matrixes were switched, and the process was repeated. The final network architecture used five historical data points – giving five input layer neurons, and four hidden layer neurons. This topology was chosen after parametric studies were run on the forecasting system.

When the system model is nonlinear and the control actions are limited in rate and magnitude, an iterative approach to predictive control can be utilized. It is important to note that when predictive control is coupled with forecasts of the system environment, antici-patory control is achieved.

The specific implementation of the predictive control system is like that described by Fieno with some exceptions. The method is easiest to explain using Figures 5.15 and 5.16. The figures show the control actions $u(t)$ and the process output $y(t)$ for an arbitrary plant during specific time frames.

In Figure 5.15 the system is repeatedly simulated from 0 to 50 seconds. During these simulations control actions 1 and 2 occurring at times 25 and 50 seconds are the variables to be optimized. Con-trol action 0 is not a variable control parameter. Once the optimiza-tion procedure has converged, that is, the integral error for $y(t)$ has reached a minimum, the system is incremented to the next time win-dow. This is the new state shown in Figure 5.16. The simulation (or control optimization) window is now 25 to 75 seconds, control action 1 is taken as the new control action 0, and control actions 1 and 2 at times 50 and 75 seconds are the new variables to be optimized over the new horizon.

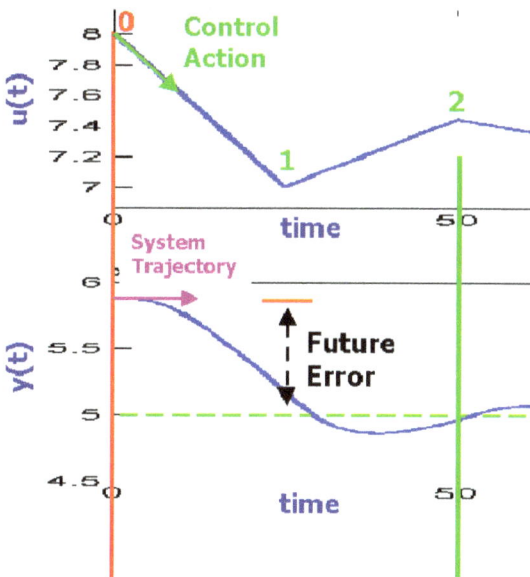

Figure 5.15. First iterative control optimization.

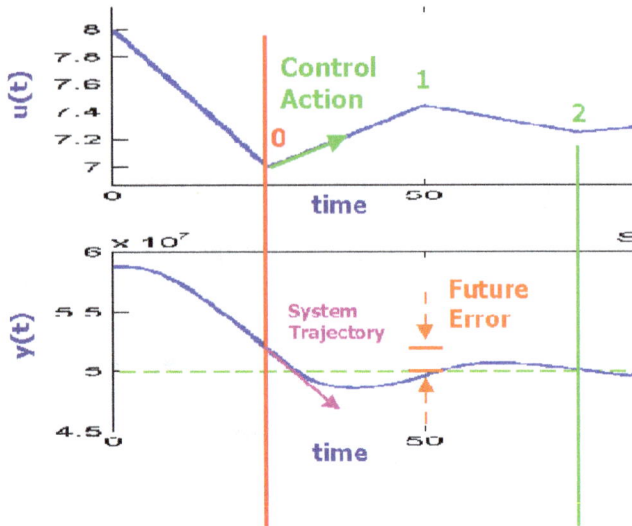

Figure 5.16. Second iterative control optimization.

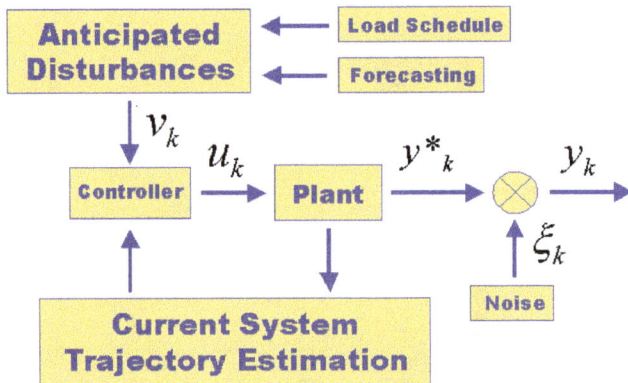

Figure 5.17. Anticipatory control architecture.

The optimization of control actions can be thought of as looking at the projected error of the system based on forecasting and any load schedules that are known *a priori*, looking at the current trajectory that the control actions up to this point are driving the system, and finding open-loop control actions to correct for the projected error. A schematic of this is shown in Figure 5.17. Note that the only feedback in the control schematic is the internal trajectory estimations of the plant.

Computational burden becomes an issue if very fast anticipatory control response is desired or if the system is too complex to simulate fast enough. A method for incorporating neural networks to alleviate this problem is presented.

When the optimal control pattern for a given set of time horizons and system states is found, the results are logged in a control action table. Figures 5.15 and 5.16 show the variables that are logged for two optimized control iterations. The predicted error for the system at the first control action is stored along with the derivative of the plant output at the current time. This allows for a control log that lists what optimal control actions to take for a given set of error circumstances.

Once the log has been created for enough examples, the relationship between the control actions and the error and system states can be learned. A neural network can be trained to accomplish this task, and this is the method used for this research project. There are certainly other ways to do this, but neural networks can model nonlinear functions very well and are computationally efficient after training (Tsoukalas and Uhrig, 1997). The neural net design used for this portion of the work has four hidden nodes, two inputs, and the same activation functions as the forecasting neural network.

The boiler model was first iterated to find suitable anticipatory control actions for a large set of know preview errors and then a neural network was trained to learn the anticipatory control relationship. Figure 5.18 shows the general response of the full-scale model to known load changes. The controller is tuned very well for this system because all the generation lines intersect near the centers of the set point step changes.

One of the advantages of anticipatory control is the ability to allow for smoother control actions than feedback control (resulting in less entropy production). This makes sense when considering the control system because the controlled variable is the change in fuel flow rate to the boiler in kg/s^2. This value is integrated over time to find the mass flow rate into the boiler system. If there is a larger time allowed for the rate adjustment, the value need not be as high to generate the same integrated signal.

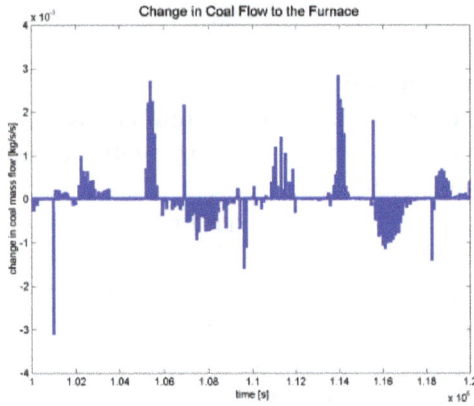

(a)

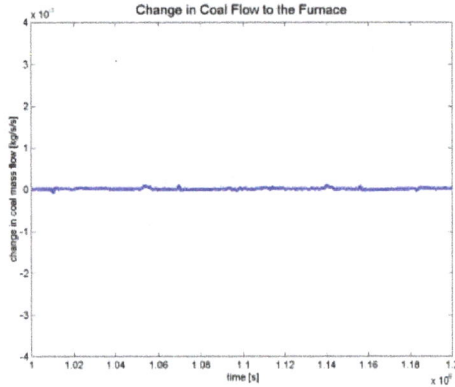

(b)

Figure 5.18. (a) There is no anticipatory strategy at the top, while in (b) there is anticipatory strategy and hence the control effort is smooth with minimal entropy production (represented by the area under the curve).

When the change in fuel flow rate is compared for a portion of the simulation case, the anticipatory control effort is about one twenty fifth of the feedback control actions. In other words, the anticipatory strategy gives better energy efficiency for *less* control effort. This reduction of control effort has the additional benefits of reducing the wear on the control system and reducing temperature gradient stresses.

Anticipatory control attempts to improve system performance by using knowledge about future events. Anticipatory control results in smoother and reduced control efforts because the system has a longer time to attempt to compensate for disturbances. This control technique, in general, is especially advantageous for systems with dead times and limited control actions. There is, however, very little previous research that ties predictive control and time series forecasting together in a unified system.

The system developed in this research is an extremely general control approach that uses very little system information and parameters in the design process. Nearly the entire design process was given to the computer to determine. The drawback to limiting the amount of information used in the controller design process is that the computational burden increases dramatically. Nevertheless, computers today are more than equipped to handle the burden.

It should be noted that the results listed are biased *towards* feedback control in two ways. The first bias is the fact that no deadtime between control signal and action was incorporated into the model. Anticipatory control would eliminate any error due to a nonlinear deadtime in the system. The second way that the results are biased is that the limits on the control actions were never achieved for either control system. If the rate limit for the feedback controller were a limiting factor, the results would have been even more favorable for the anticipatory control strategy.

The anticipatory control system presented is illustrative. There are many possible improvements and possible applications for this work. It would be interesting to see how universally applicable the anticipatory control system would be. The controller developed should be tested on other systems that have highly nonlinear environments and are manipulating the environment with a nonlinear system.

For the specific improvement of the system developed in this research, there are two main areas where improvements are possible. The first of these areas is in the prediction of the environment or electric load of the grid. Literature shows that the use of weather data in electric load forecasting can make the predictions more accurate. This makes sense physically because on very hot days the electric load is higher due to increased usage of air conditioners. It is recommended that weather information be tested for inclusion in the environmental prediction scheme.

Another possible way to improve the prediction system is to decompose the time series into long-term trends and short-term trends. This would allow capturing seasonal variations in power demand separately from weekly variations.

The second area where improvements in the model are possible is in the predictive control of the power plant. The controller designed for this project is optimal for this specific application but if a change occurs that affects the time dynamics of the plant, the performance of the tuned controller will suffer greatly. An adaptable tuning mechanism for changes in the plant dynamics using a system identification approach would allow the plant controller to be adaptable – not just the environment forecasting system.

Anticipation of future states based on current and recent observations is crucial for tasks such as autonomous operation, fault detection, and predictive maintenance in complex energy production systems, including Generation IV (GenIV) nuclear reactors. While traditional statistical modeling is sometimes useful, it often fails to capture the nonlinear behavior of these systems. Deep learning architectures have emerged as promising alternatives due to their ability to learn without remembering; that is, to generalize patterns from historical data without explicitly storing the full history. This capacity makes them well-suited for anticipatory tasks in reactors, particularly in forecasting sensor time series such as coolant temperature, pressure, or flow. Once they are trained, ML models do not recall specific training examples. Instead, they internalize statistical relationships and dynamics through optimized weight parameters. This allows them to anticipate future system states even when past observations are no longer explicitly available.

Example 5.7. Prediction with Long Short-Term Memory (LSTM) Networks.

Long Short-Term Memory (LSTM) networks were originally developed to overcome the limitations of traditional recurrent neural networks (RNNs), particularly the vanishing gradient problem.

In energy systems, LSTMs have proven effective in forecasting time series data by selectively retaining relevant temporal features through their gating mechanisms.

Each LSTM *cell* contains an *input* gate, *forget* gate, and *output* gate, that control the flow of information across time steps. For each time step, the LSTM cell calculates hidden state H_t and current state C_t according to:

$$H_t = \sigma(x_t U_o + H_{t-1} W_o) \tanh(C_t)$$

$$C_t = \sigma(x_t U_f + H_{t-1} W_f) C_{t-1}$$
$$+ \sigma(x_t U_i + H_{t-1} W_i) \tanh(x_t U_c + H_{t-1} W_c)$$

where

x is the input to the cell,
U and W represent weight matrices, and,
i, o, f, are indices referring to *input*, *output* and *forget* gates, respectively.

Figure 5.19 illustrates the architecture of an *LSTM cell* which is the core functional unit responsible for managing memory, updating states, and passing information forward through time steps. It contains gates (input, forget, and output) which regulate the flow of information, enabling capture of long-term dependencies.

The ability of LSTMs to forecast stems from patterns formed during training. As an example, in a study investigating time series forecasting in Gen IV thermal hydraulic systems, LSTMs were pre-trained on low-temperature surrogate fluid data and successfully fine-tuned with minimal high-temperature data to anticipate future temperature of the coolant (Pantopoulou *et al.*, 2025).

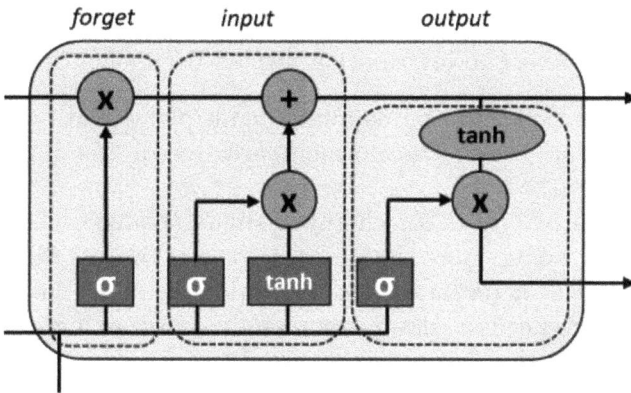

Figure 5.19. LSTM cell is the core functional unit of Long Short-Term Memory networks used in deep machine learning.

Their forecasting accuracy was highest when using a moderate lookback window (around 20 time-steps), indicating that overly long historical windows did not enhance, and even degraded the performance due to overfitting or redundant inputs.

Example 5.8. Transformers in Anticipation.

Transformers were originally developed for natural language processing. They use the attention mechanism to capture temporal dependencies. Transformers extend traditional convolutional neural networks (CNN) by incorporating transformation pooling mechanisms, allowing learning and generalization across various transformation groups without relying on predefined parameters. Transformers do not process data sequentially, but instead evaluate the relevance of all past inputs in parallel through learned attention weights. This structure enables them to focus on critical segments of the input history when making predictions, regardless of their temporal position. The attention mechanism creates weights that measure the relevance between parts of the input. Similarity scores are calculated from the relationships between the query (Q), i.e., the current data point, and each key (K), i.e., all other data points. The attention function is calculated as

$$\text{attention}(Q, K, V) = \text{softmax}\left(\frac{QK^T}{\sqrt{d_k}}\right)V$$

where

d_k is a scaling factor, and
V (value) contains the information needed from each data point.

This mechanism can be applied multiple times (multi-head attention). The basic Transformers architecture is illustrated in Figure 5.20.

In anticipatory tasks for energy systems, Transformers have the advantage of being more stable over longer forecast horizons. As demonstrated in a recent study (Pantopoulou *et al.*, 2025). Transformers experienced a slower increase in forecasting errors with increasing prediction horizons compared to LSTMs.

While LSTMs present higher short-term accuracy and typically converge faster due to fewer trainable parameters involved, Transformers offer slower performance degradation as the forecast horizon

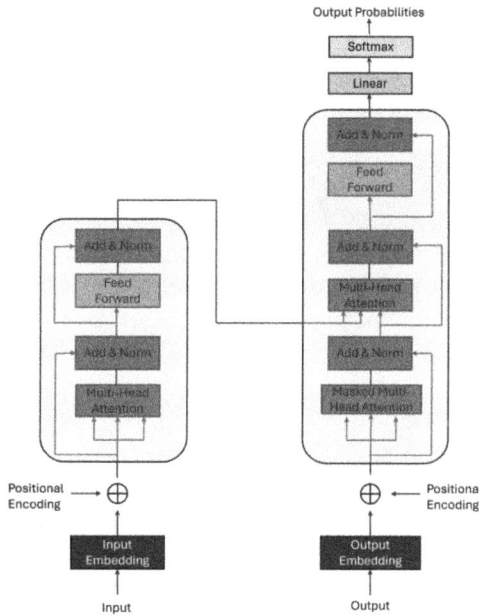

Figure 5.20. Architecture of transformers.

increases, thus performing better in long-range forecasting tasks. In the mentioned study, while both models achieved their best accuracy with lookback windows in the vicinity of 20 points, Transformers errors increased less rapidly as the window lengths increased. This suggests that hybrid architectures or ensemble methods that combine the short-term precision of LSTMs with the long-term generalization of Transformers might be a viable solution to automate anticipatory tasks in these systems.

5.6 Generation-Consumption Networks and Anticipation

Modeling energy as a smart object is a novel and significant step towards smart energy systems but with the assumption that an object should have a fairly long-life cycle to allow interactions (talk, trade, negotiate, etc.) to happen. For electrical energy this does not hold, since physical generation and consumption of electricity are almost simultaneous which forbids any meaningful actions to take place in between. Therefore, storage is critical, especially for a peak demand constraint system. In the following, we illustrate that with a simplified resource generation and consumption network.

A simplified resource generation-and-consumption network is shown in Figure 5.21. Assume for simplicity that the delivery system is endowed with unlimited capabilities. All three components (generation, storage, and consumption) are considered dynamic and can be characterized by their state functions. A peak demand constraint requires that the following condition be met at all time.

$$g(t) + s(t) > c(t) \tag{5.19}$$

It should be noted that these three components are time-varying with different time constants. The quantity $c(t)$ in Eq. (5.19) represents customer behavior. In the absence of a feedback loop, customer behavior is decoupled from or independent of the generation side.

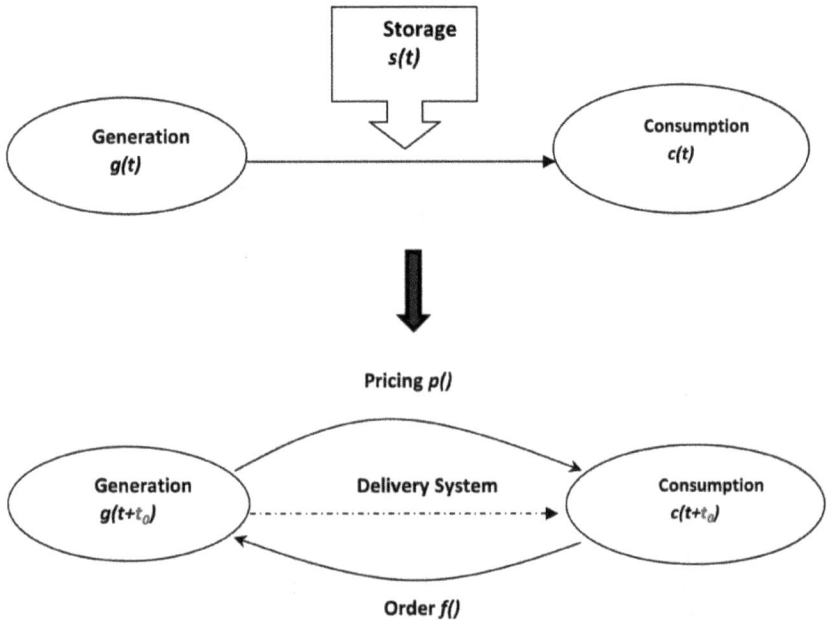

Figure 5.21. A simplified (open system) generation-consumption network (top) transformed into a closed system with anticipation.

It has a relatively small time constant, which implies that it could change abruptly. The quantity $g(t)$ in Eq. (5.19) represents generation capacity. The change of generation causes a very large time constant. The quantity $s(t)$ in Eq. (5.19) represents storage capacity. Depending on the storage technology, its time constant varies. In the electric power network, large scale storage of electricity via battery or chemical processes is not yet technically and financially feasible. Therefore, it is up to the generation side to keep the safety margin mostly through so-called spinning reserve, which is a major source of energy waste.

A careful examination of Figure 5.21 reveals some important issues. First, it operates at a single time scale which is the current time, in which everything runs in real-time. Second, there is no feedback from the customer side; therefore, balancing must be achieved in a unidirectional manner. Since $c(t)$ is independent of $g(t)$ and $g(t)$ is relatively slow to change, $g(t)$ must stay at a very high level to maintain the inequality. While there is no feasible solution for large scale energy storage, we propose to implement a "virtual" storage through anticipatory demand-side management. With the emergence of smart energy management tools, which control multiple smart appliances or smart devices, it is possible to encode a customer's consuming pattern into devices (McDonald and Thomas, 2021).

By anticipating future energy demand and future states of an energy network (load, disturbance, etc.), energy behaviors with powerful conservation and environmental benefits can be optioned and incentivized into any power-consuming device. It is an effective regulatory tool to utilize short-term elasticity along with pricing information.

Anticipation and short-term elasticity are two key elements for a smart energy system. The synergism of the two enables energy savings in a smart energy system, which eventually leads to a more stable and sustainable infrastructure. This can be illustrated with the aid of an example. Figure 5.22 shows a typical power consumption pattern over 120 hours. As stated before, a smart energy system will have to be implemented through a smart energy object. We see in the following analysis that a system's smartness largely depends on the way the energy object is defined.

First, let's start with a least smart object, which means the smallest number of attributes are defined or the object has the lowest

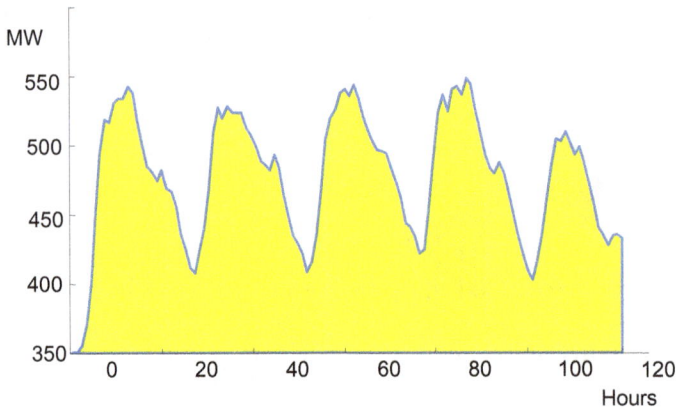

Figure 5.22. A typical five-day power consumption pattern in a LAG.

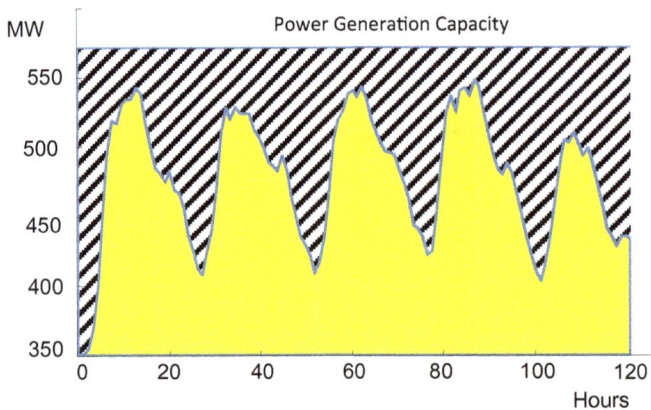

Figure 5.23. Fixed power generation meets peak demand (plus safety margins) but wastes a lot of energy (shaded area).

dimensionality. Let's assume that there is no anticipation information available for the power consumption behavior. A simple and safe plan is to maintain a fixed and high level of generation capacity by adding a safety margin to any estimated peak demand, as shown in Figure 5.21. This, however, causes a lot of waste due to the unused capacity, which is indicated by the shaded area in Figure 5.23.

Next, we make the system smarter by including an additional dimension into the energy object, which is the forecasted consumption. If the power consumption can be forecasted to some degree of

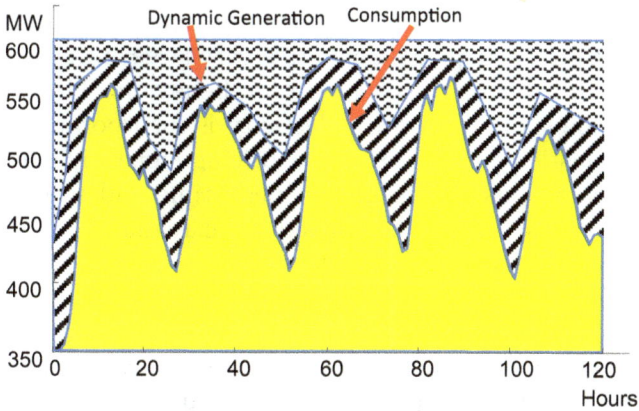

Figure 5.24. Dynamic generation with anticipation can meet peak demand (plus safety margins) while significantly reducing energy waste.

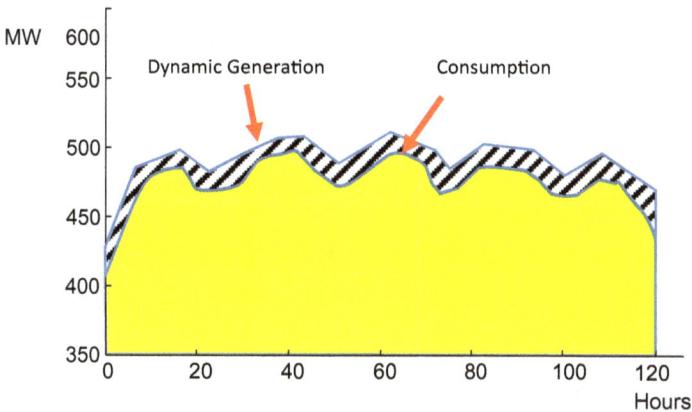

Figure 5.25. Dynamic generation with anticipation and short-term elasticity (demand response) can minimize energy waste.

accuracy, it is then possible to schedule the power generation dynamically to follow the actual consumption, as shown in Figure 5.25. The forecasting model cannot be perfect, and for safety reasons a margin must be added to the prediction, which leads to the energy waste indicated by the shaded area in the middle part of Figure 5.24. But it already is an improvement.

We can further increase "smartness" of our model by adding more dimensions, such as short-term elasticity. Anticipation provides an energy savings tool by generating energy only when it is needed. However, the forecasting could lead to a very large error for some part of the time due to irregular consumption behaviors. This is unavoidable unless the short-term price elasticity is incorporated. With the short-term elasticity tool, through a pricing signal, it is possible to discourage energy usage during the peak time and to encourage it during the off-peak time. As a result, the regularity of the power consumption pattern improves. Accordingly, generation can follow consumption more closely, as shown in Figures 5.23 and 5.24.

The process can continue by adding such info as routing paths. Using this attribute we can reduce traffic, avoid overloading on some critical parts of the power grid, and ultimately improve reliability. There are several things we learn from this example:

1. Defining energy as objects is a critical step towards a smart energy system. The degree of smartness is determined by the complexity of the object. The more attributes are defined, the smarter the system can be.
2. Smart energy objects need to be instantiated and operated by smart devices.
3. For smart electricity object to be reliable, virtual buffer support is required. Virtual buffer is achieved through pricing and negotiation and exists as a form of contracts and agreements. Virtual buffer requires that smart energy systems operate with anticipation.

Using objects is an effective approach to model a complex energy system. The model represents a virtual energy system; its usefulness depends on how it correlates to the real physical system. In this section, we examine the relation between the two systems with state space concepts.

One essential element defining an energy object is the temporal dimension, including properties such as current time, scheduled production time, scheduled consumption time, etc. With properties on different time scales, objects can model future operations in the world through anticipations. Anticipations are projections about future and possible states of a system. The accuracy of anticipations determines how close a model is to the physical system and there are always prediction errors.

Therefore, it is very rare in the real world that a predictive model is used alone. Instead, the model relates to the physical system in a closed loop so that feedback or observations from the physical system can be used to reduce prediction errors. However, this approach does not apply to energy systems, especially the electric power grid. In a conventional feedback control system, there is an often-overlooked assumption that the model must run faster than the physical system.

Only with this condition has the model sufficient time to perform prediction and correction tasks. In an electric power system, the object – electricity – travels nearly at the speed of light; there is no computational tool to surpass that. Hence, as shown in Figure 5.26 the model and real system must be represented in two separate spaces: The projected state space and the physical state space.

The distinction of the two spaces is enforced by physical laws; we show next that an association can be made artificially. Represented in the physical state space, a complex energy system is a non-steady state and non-equilibrium system. Modeling and managing such a system autonomously is problematic mathematically and numerically. A complex system is known to be plagued by the instability problem; once a system is trapped into an unstable state cascading failures may follow. And it is nearly impossible to identify all unstable regions ahead of time to avoid them.

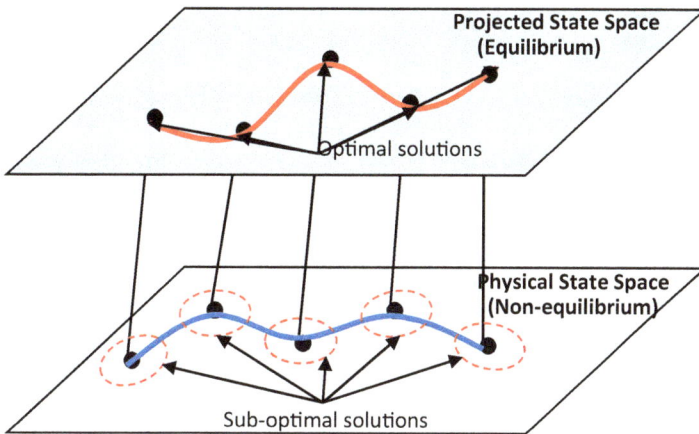

Figure 5.26. Augmented state space.

On the other hand, in the projected state (virtual) space, a model can be optimized, to avoid this problem. Since only virtual not physical objects are involved, it is possible to bypass the physical limitations and achieve equilibrium with optimal solutions. In a smart energy system, the optimal solutions will be sought through negotiations among all parties each of which is represented by an intelligent agent. Given a good strategy, an agreement that maximizes the overall benefits can be reached within a reasonable time frame. The optimal solutions in the virtual space will not necessarily turn to reality in the physical state space. However, the contractual commitment will act as an adequate force to confine the physical system to at least some suboptimal solutions (Tsoukalas and Gao, 2008a, 2008b).

5.7   Indicators, Predictors and Data Brevity

The grid-edge has become the locus of ongoing expansion of sensors and effectors which was previously centralized under the Supervisory Control and Data Acquisition (SCADA) systems. At the grid edge, intelligence is shifting towards distributed automation. Instead of relying solely on a central SCADA system, intelligent edge devices can detect anomalies and respond autonomously within milliseconds – sometimes even before the central SCADA system registers an event. This transformation is driven by the increasing complexity of modern power grids, which integrate renewable energy sources, electric vehicles, and smart loads. Edge computing enables real-time decision-making at the grid edge, improving stability, efficiency, and resilience. he overall electric power system, the list of objectives can grow quickly before we realize that there is no ideal existing solution to satisfy all the expectations, some of which are even conflicting in nature. For example, making more profits for utilities is usually achieved by maximizing sales, which contradicts resource conservation or emissions goals. Therefore, a promising smart solution should be able to prioritize objectives, resolve conflicts and objectively evaluate options. A principal challenge has to do with the enormous volume of data. Digitization is great for AI solutions, on the other hand, unless the brevity of data is addressed, even AI may have difficult time to track things over long periods of time. It is imperative therefore to extract indicators and features from the data, to manage observables that represent important, information reach signatures and approach the data brevity problems in a multi-sensor, multi-temporal, and multi-spatial approaches. All these features are considered difficult tasks to be dealt with in an autonomous way especially at agents controlling high level intelligent entities, that is, smart devices, systems, and processes in the grid-edge. On possibility is to use transforms such as Hilbert-Huang, Fourier, or Wavelets. The last one can be thought of as an extension of the traditional Fourier transform. In the Fourier transform, a signal is decomposed into a series of sine waves as shown in Eq. (5.20).

$$X(\omega) = \int_{-\infty}^{+\infty} x(t)e^{-j\omega t}dt \tag{5.20}$$

The analyzed signal $x(t)$ is viewed a weighted sum of a series of base functions. Sinusoid wave is an oscillatory function in time domain, spreading from $-\infty$ to $+\infty$ and appears to be a δ function in the frequency domain. This characteristic determines that sinusoid wave has perfect resolution on frequency domain while contains no temporal information in the time domain. This is undesirable for many feature-base applications where the features (usually the singularities) in a signal are required not only to be detected but also to be located temporally. Traditional Fourier transform is not good at such applications because its base function is too "long" to carry any temporal information. One intuitive modification is to employ a shorter base function to define wavelet transform (WT) as shown in Eq. (5.21).

$$WT_x(a, \tau) = \frac{1}{\sqrt{a}} \int x(t)\psi^* \left(\frac{t - \tau}{a}\right) dt \tag{5.21}$$

where $\psi(\cdot)$ is called mother wavelet, which at least satisfies that it is finite supported in time domain. The parameters τ and a are translation factor and scaling factor respectively. Some candidates form mother wavelets are shown in Figure 5.27. Because $\psi(\cdot)$ is finite supported in time domain, it is qualified to convey temporal information. Due to constrains by the Uncertainty Principle, while wavelet analysis gains resolution in time domain it must lose some resolution in frequency domain.

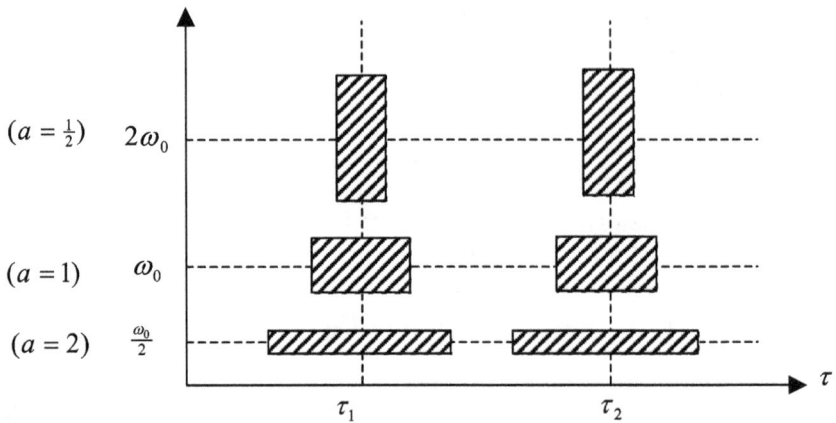

Figure 5.27. Time-frequency space of wavelet functions.

The wavelet transform decomposes a signal into a weighted sum of a series of *wavelet functions*. The weights are called *wavelet coefficients*. Wavelet functions are obtained by *translating* and *dilating* a *mother wavelet function*. At first glance there is not much new compared to the Fourier transform expect for the introduction of two new parameters: *scaling* and *translation* factors. These two factors are necessary to make the wavelet transform complete because wavelet functions are finite supported both in time and frequency domain. Although there appears to exist a simple and natural step from Fourier to wavelet transforms, there exist some noteworthy and important differences:

- The wavelet transform can characterize local features. The wavelet coefficient is a function of translation factor, thus temporal information is kept within the wavelet coefficients.
- The wavelet transform is a multi-resolution (or multi-scale) analysis. The wavelet coefficient is also a function of the scaling factor. From the viewpoint of filtering, wavelet functions are like band-pass filters with different center frequencies and bandwidths determined by the scaling factor. Thus, the wavelet transform is equivalent to filtering a signal using a series of band pass filters. The wavelet function corresponding to a small scaling factor a has narrow support in time domain but wide support in frequency

domain. This wavelet function then has coarse resolution in frequency domain. On the contrary, a large scaling factor generates a wavelet function having wide support in the time domain but narrow support in the frequency domain. Such a wavelet function then has relatively high resolution in frequency domain. This property is illustrated in Figure 5.31. Time-frequency space of different wavelet functions is shown in Figure 5.31. By selecting different scales, a wavelet transform is analyzing the signal with different resolutions. The resolution of time-space features can be simulated by the behavior of human being. If he/she wants to gather an overall image of an object, he/she will use a coarse resolution with big view angle to observe. Suppose he/she would like to focus on some details, he/she then needs to move closer to the object to obtain a finer resolution.

- The *quality* of wavelet functions is a constant. The quality of a band-pass filter is defined as the value of its central frequency divided by its bandwidth. As clearly shown in Figure 5.31, as the central frequency increases, the bandwidth increases proportionally. Thus, the quotient remains constant.

Wavelet's versatility can be alternatively exemplified by the multiple choices possible for the mother wavelet. It is very convenient to be able to select an appropriate mother wavelet to meet the practical requirements of a particular problem. However, such kind of selection is not arbitrary; reconstruction calls for information found in the wavelet coefficients. This implies that not any arbitrary function is qualified as a mother wavelet. It can be shown that only those functions that meet certain *admissibility condition* allow their inverse wavelet to transform to perfectly reconstruct the original signal from the wavelet coefficients. The *admissibility condition* is stated in Eq. (5.22).

$$c_\psi = \int_0^\infty \frac{|\Psi(\omega)|^2}{\omega} d\omega < \infty \qquad (5.22)$$

In Eq. (5.6), $\Psi(\omega)$ is the Fourier transform of a mother wavelet. Then the original signal can be reconstructed as shown in Eq. (5.22).

$$x(t) = \frac{1}{c_\psi} \int_0^\infty \frac{da}{a^2} \int_{-\infty}^{+\infty} WT_x(a, \tau) \frac{1}{\sqrt{a}} \psi\left(\frac{t - \tau}{a}\right) d\tau \qquad (5.23)$$

The wavelet transform contains a continuous time and a continuous scale factor and consequently it is called *continuous wavelet
transform*. By discretizing the translation factors (time) and scale
factors (frequency) we can obtain a digital version of the continuous wavelet transform: discrete wavelet transform, which is shown in
Eq. (5.24).

$$DWT_x(j,k) = \sum_n x(n)2^{-j/2}g(2^{-j}n - k) \qquad (5.24)$$

In Eq. (5.24) a new symbol $g(\cdot)$, called wavelet filter, is used in
place of $\psi(\cdot)$ is ordered to make wavelet transform look more like a
filtering operation. Wavelet filters perform an analogous role to that
of the wavelet functions found in the continuous wavelet transform.

The wavelet transform appears to be a concept, which is far away
from artificial neural networks. But upon scrutiny, one finds that
the wavelet transform is a multi-resolution technique the rudiments
of which may be found in the learning behavior of humans. This
suggests the possibility of neural networks improving their learning
abilities by exploiting the power of the wavelet transform. A characteristic of the wavelet transforms that neural network could also
utilize is its superb performance on singularity detection. The singularity of a signal is often referred as *discontinuity, rapid change,
transient* or *irregularity*. The singularity part in a signal is thought
to contain more information than elsewhere. A major task of neural networks is to collect information from all singularities in given
samples and try to extract appropriate patterns to classify them. In
this case, the wavelet transform can serve a preprocessor for, or an
integrated component of, neural networks.

Two major limitations of back-propagation leaning algorithms are
well known: *convergence rate* and the *local minima problem*. If these
two problems are not handled properly, either a solution may be
obtained after a very long period of training or an optimal solution
may never be found. Both scenarios are unacceptable for practical
applications, especially for highly nonlinear problems where a neural
network is more likely to be "stuck" in a local minimum. Because
little or no *a priori* knowledge about the problem is utilized in a
neural network, the search space must be very large to find a solution.
The bigger the search space is the slower the convergence rate and

the more likely it becomes to meet a local minimum. An efficient approach to alleviate this dilemma is to narrow the search space.

Wavelets are an excellent tool for narrowing the search space of neural networks. Consider, for example, what is involved when a person identifies an object. It is almost impossible for anyone to randomly grasp a piece of information from the object, match it with a database and decide what the object is (this is what pattern recognition programs do). A reasonable way for the person to follow is first to use a very coarse resolution to observe the object and gain an overall profile that serves one of possibly several premises for his final decision. Next, he probably will increase the resolution and focus on some local features. The integration of all information from these two channels leads to the final decision. This simple example illustrates how the pattern recognition process used by humans is a multi-resolution approach. Suppose a person is a constant-resolution observer. If the resolution is too coarse, only a profile can be observed, and he will not be able to gather enough information to distinguish it from others. In this case, he might not be able to discriminate a football from a coconut because they share a similar shape. Going to the other limiting extreme, if the resolution is too fine, he is very likely to be overwhelmed by the explosion of information because he must deal with some many details. Similar examples can be found in the employment of neural networks for modeling and forecasting purposes.

There is an intuitive connection between wavelets and human intelligence. Some biological evidence has been recently found indicating that biological (not artificial) neurons are performing wavelet computing in the brain. The evidence was presented by S. P. Dear and C. B. Hart in 1999. It involves recording the cortical extracellular multiunit potentials from two adult bats that were exposed to a sequence of acoustic stimulus that are typical of bat insect pursuit and capture. The researchers have discovered that the waveform of averaged potentials could be classified that very resembling the waveform of 17 members of the Symlet wavelet packet family. More surprisingly two important wavelet characteristics, dyadic scaling, and dyadic translation, are also found in these biological signals.

Using wavelets as an integral part of neural networks was presented by Zhang and Benveniste in 1992. For simplicity, let's consider a single input single output case. The key point is that generally a neural network can approximate any function. This function, from the viewpoint of wavelet transform, can be represented by a series of wavelet functions. Let's assume that N of these wavelet functions can cover the time-frequency space of the signal effectively (acceptable in terms of approximation). Then this signal is represented using these N wavelet functions as Eq. (5.7):

$$\hat{f}(x) = \sum_{i=1}^{N} c_i \psi(a_i x - b_i) = \sum_{i=1}^{N} c_i \psi_i(x) \qquad (5.25)$$

In Eq. (5.9) variable x has the same meaning as time. The original signal $f(x)$ is approximately by a linear combination of wavelet functions. The linear coefficients c_i can be treated as connections weight in a neural network. A single input neural wavelet network is shown in Figure 5.28.

The output of the network shown in Figure 5.19 is expressed as Eq. (5.26):

$$f(x) = \sigma \left(\sum_{i=1}^{N} w_i \psi_i(x) \right) \qquad (5.26)$$

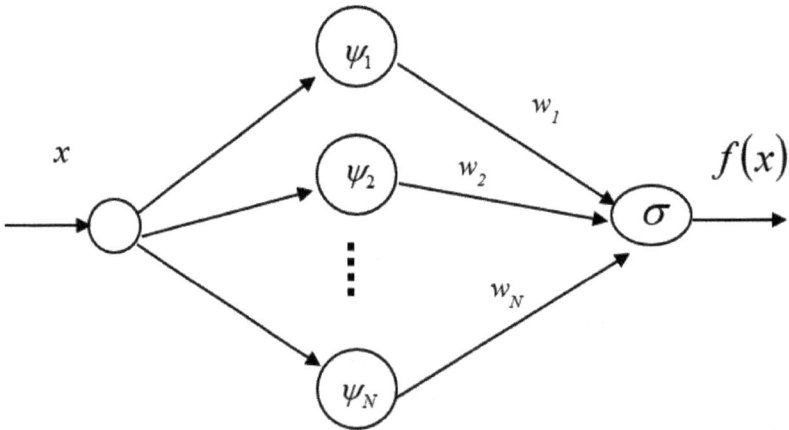

$$f(x) = \sigma \left(\sum_{i=1}^{N} w_i \psi_i(x) \right)$$

Figure 5.28. Neural-wavelet mid-term forecasting architecture.

where $\sigma(\cdot)$ is the activation function of output neurons. Since the adjustable weights appear only at the output layer, any standard learning algorithm can be used for training.

Usually, dyadic wavelets are used as basis functions, which are generated by Eq. (5.27):

$$\psi_{j,k}(x) = 2^{-j/2}\psi(2^{-j}x - k) \qquad (5.27)$$

In this situation, following stable condition has to be satisfied to reconstruct the function f

$$A\|f\|^2 \le \sum |\langle f, \psi_{j,k}\rangle|^2 \le B\|f\|^2 \qquad (5.28)$$

where $0 \le A \le B < \infty$.

The single input neural wavelet shown in Figure 5.23 can be interpreted as using a wavelet synapse in an ordinary neural network. It does not seem to use much power of wavelets whereas a multi-input neural wavelet provides a better approach.

In a multi-input neural wavelet network, every neuron in the hidden layer is functioning as a wavelet filter.

The neural wavelets we have discussed so far (single-input or multi-input) have adjustable weights which appear only at the output layer and all other parameters (such as the scaling and translation factors for wavelet functions) are fixed. This arrangement facilitates the design of a learning algorithm. We only need to consider the updates of output weights, which is rather easy to do. Yet, the small number of adjustable parameters may limit the performance of the neural network because the solution space spanned by these parameters may not contain the optimal solution. To overcome this problem, Kobayashi and Torioka presented a modified neural wavelet. Two major changes in their design are identified:

- The number of hidden neurons varies, which implies that the number of wavelet filters is determined dynamically.
- The scaling and translation factors of each wavelet function are adjustable. A learning algorithm is applied to select an optimal configuration for each wavelet filter so that these filters cover the time-frequency space of the input signal.

The increase in the numbers of adjustable parameters undoubtedly enhances the flexibility of neural wavelets. However, it complicates the design of a learning algorithm.

As discussed earlier, an *admissibility condition* is needed to enable the computation of the wavelet coefficients. It seems that this rule should be strictly obeyed in choosing a mother wavelet for neural wavelet network. Yet, this is not true as Yamakawa has shown in 1994. The idea is that the wavelet coefficients in a neural wavelet network are not obtained through computing but through learning. Thus, the admissibility condition is not necessary to follow, which provides us much freedom in designing a neural wavelet network. Furthermore, the author claims that the traditional choice of an orthogonal wavelet bases is inappropriate under some circumstances. Because "an orthogonal basis function of wavelets exhibits complications," the output may be wavy or oscillatory, which is a serious problem for such applications as system identification and chaotic system prediction. An alternative set of wavelet functions, called *over-complete number of compactly supported non-orthogonal wavelets*, was presented, as shown in Eq. (5.29) and Figure 5.29.

$$\psi(x) = \begin{cases} \cos(\pi x) & (-0.5 \le x \le 0.5) \\ 0 & \text{otherwise} \end{cases} \tag{5.29}$$

$$\psi_{a,b}(x) = \psi(ax - b)$$

The wavelet bases are not orthogonal but simple and smooth, which are supposed to produce more stable results. Moreover, over-complete number of bases are adopted to obtain improved generality. It is shown that this structure gives better performance in predicting the chaotic behavior of nonlinear systems.

Neural network integrated with wavelet elements is a revolutionary idea, which makes neural networks smarter in learning. Many applications of neural wavelet network in function approximation, channel equalization, optical character recognition, nonlinear prediction, spectral compression, classification, and system identification already shown that it convergent much faster than an ordinary feed forward network.

The introduction of wavelets into neural networks also changes a fundamental characteristic of neural networks. The neurons in a neural wavelet network are no longer identical, they have different

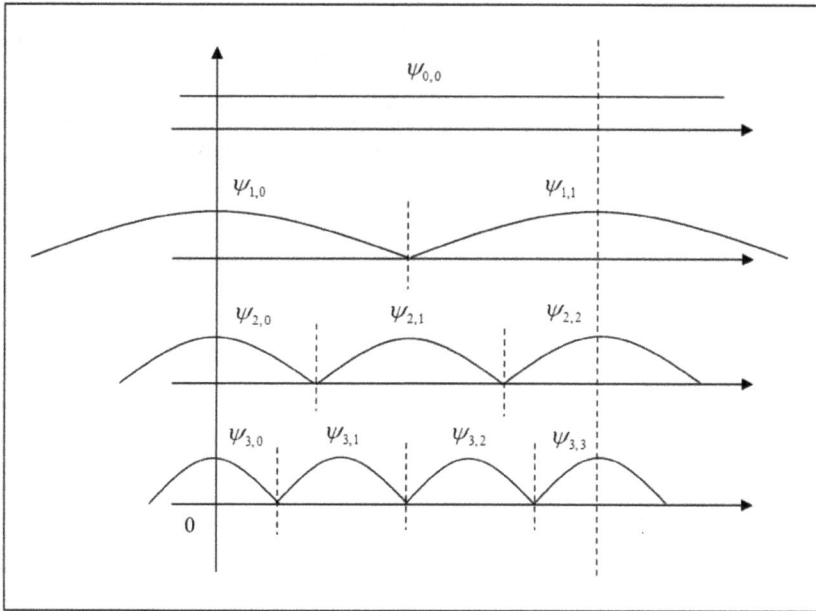

Figure 5.29. A family of non-orthogonal basis functions.

resolutions and different scopes. This is a significant step toward a real diversified system, which we deemed to be a unique strength that belongs to biological systems. Conceptually we can image that a diversified system can operate more reliably and successfully.

The current model of neural network does not work well to predict the mid-term or long-term power loads. The problem is that it fails to utilize some information that is unique or idiosyncratic to individual power loads. It is a fact that the power consumption at a specific time in a given day highly correlates not only to those at previous hours in the same day but also to those at the same time in previous days, as shown in Figure 5.30.

Suppose the current time is noon in Day 3. Then it is naturally that the current demand has a close relationship with the power loads at 11 AM and 10 PM in Day 3 because they are in the same day. Moreover, the current demand is also similar to the power consumptions at same time of Day 2 and Day 1 because they are in the same time. This dual temporal correlation is to be exploited in our next step and can easily be accomplished by rearranging the data block.

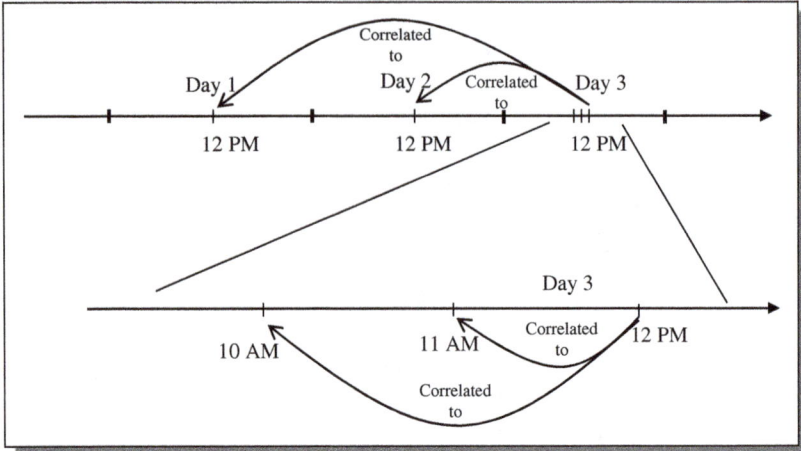

Figure 5.30. Dual temporal correlation and anticipatory model for power demand (Gao and Tsoukalas, 2007).

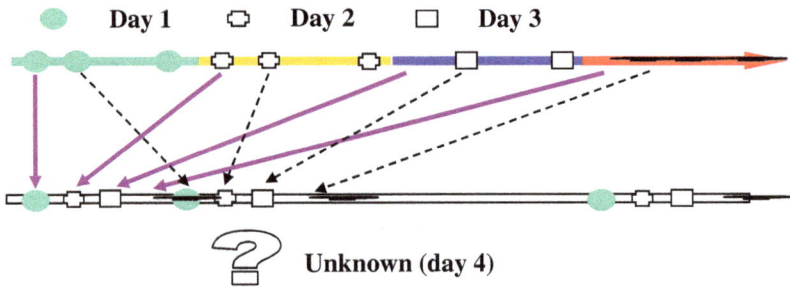

Figure 5.31. Data analytics for neural wavelet predictor.

An interleaved arrangement of data will be used. We could make a model for short term or even long-term prediction.

Neural wavelets have been very successful as function approximation, which is a data interpolation problem. In prediction, we need to do extrapolation instead. Thus we need to manipulate the data in an anticipatory way so that an extrapolation problem can be solved by interpolating.

Figure 5.31 exemplifies the basic idea involved. The resultant data compound is not organized in increasing time base (as is usually done) where the historical data (known) are put in the beginning and the data to be predicted (unknown) at the end. Instead, we mix the known and unknown data in such a way that the dual temporal correlation is exploited. In Figure 5.31, the data to be predicted are represented by question marks. The historical data from three days before, two days before and one day before are represented by circles, crosses, and squares respectively. The bottom part of Figure 5.31 shows the way of mixing these data. Generally, to predict the load at a certain time, we use the load of three previous days at this time. Moreover, the historical data at the next time step is utilized as the anticipatory purpose. This arrangement is an effective way to avoid over-fitting.

An example of a 24-hours-ahead prediction is shown in Figure 5.25. As can be easily seen by inspecting this figure, the results are very promising. The prediction follows the overall trend of the actual data and no observable over-fitting problems are present. The major error appears in the small load region (lower left side). This can be explained and attributed to the use of mean square error as driving function inside the neural network. Other alternatives such as relative percentage error may improve it.

The power of wavelets decomposition renders the neural wavelets approach unique power in function approximation and prediction problems. By appropriately selecting the wavelets base, the over-fitting problem can be effectively avoided. Moreover, we can carefully arrange the original data to take advantages of neural wavelet approximator to perform mid-term and long-term prediction of power customer behavior.

Different approaches to decomposition can be used to fit idiosyncratic short and mid-term predictions, including fast Fourier transforms and the Hilbert-Huang transform.

Problems

Problem 5.1. Given the peak demand data of the hyper data center in Example 5.3, predict the peak demand for the 11th year through a polynomial regression.

Problem 5.2. Given the peak demand data of the hyper data center in Example 5.3, predict the peak demand for the 11th year through exponential regression.

Problem 5.3. Given the peak demand data of the hyper data center in Example 5.3, predict the peak demand for the 12th and 13th year through linear regression.

Problem 5.4. Given the peak demand data of the hyper data center in Example 5.3, predict the peak demand for the 12th and 13th year through polynomial regressions.

Problem 5.5. A small rural town is part of a LAG and has the following monthly demand:

January:	6,500 kW
February:	6,200 kW
March:	5,300 kW
April:	6,000 kW
May:	6,300 kW
June:	7,150 kW
July:	7,240 kW
August:	6,900 kW
September:	5,750 kW
October:	5,300 kW
November:	4,900 kW
December:	5,800 kW

(a) Plot the monthly average demand for the entire year.
(b) What is the annual load curve?
(c) What is the annual average demand?
(d) What is the annual load factor?

Problem 5.6.

(a) Graph theory problem, nodes and links
(b) How many links in a completely connected graph of $n = 5$ nodes, etc. (like Example 5.1)
(c) What is the number of links L for a network of nodes n (without self-referring nodes).

Problem 5.7. Consider a divisibility relation on the set of natural numbers $S = \{1, 2, 3, 4, 5, 6\}$

(a) Is it a POSET
(b) If it is a POSET, what is the Hasse diagram?

Problem 5.8. Given the sets:

$$A = \{1, 2, 3, 4, 5\}$$
$$B = \{4, 5, 6, 7\}$$

Calculate the *Jaccard Index* to find the similarity of sets A and B.

Problem 5.9. Given the sets:

$$A = \{1, 2, 3, 4, 5\}$$
$$B = \{4, 5, 6, 7\}$$

Calculate the congruence C, given by Eq. (5.2). How does it compare to the *Jaccard Index* found in Problem 5.4 and what is the interpretation about the difference?

References

Amin, M., "The smart-grid solution," *Nature*, 499, 145–147, 2013. https://doi.org/10.1038/499145a.

Gao, R. and Tsoukalas, L.H., "Implementing virtual buffer for electric power grids," *Proceedings of ICCS 2007*, Beijing, China, 2007.

McDonald, J.D. and Thomas, M.S., *Power System SCADA and Smart Grids*, CRC Press, Boca Raton, FL, 2019.

Pantopoulou, S., Cilliers, A., Tsoukalas, L.H., and Heifetz, A., "Transformers and Long short-term memory transfer learning for GenIV reactor temperature time series forecasting," *Energies*, 18, 2286, 2025. https://doi.org/10.3390/en18092286.

Prantikos, K., Chatzidakis, S., Tsoukalas, L.H., and Hiefetz, A., "Physics-informed neural network with transfer learning (TL-PINN) based on domain similarity measure for prediction of nuclear reactor transients," *Scientific Reports*, 13, 16840, 2023. https://doi.org/10.1038/s41598-023-43325-1.

Rosen, P., *Anticipatory Systems Philosophical, Mathematical & Methodological Foundations*, Pergamon Press, New York, 1985.

Tsoukalas, L.H., *Fuzzy Logic: Applications in Artificial Intelligence, Big Data and Machine Learning*, McGraw Hill, New York, 2023.

Tsoukalas, L.H. and Gao, R., "Inventing energy internet: The role of anticipation in human-centered energy distribution and utilization," *Proceedings of 2008 SICE Annual Conference*, pp. 399–403, Chofu, Japan, 2008a, doi: 10.1109/SICE.2008.4654687.

Tsoukalas, L.H. and Gao, R., "From smart grids to an energy internet: Assumptions, architectures and requirements," *Proceedings of IEEE DRPT*, Nanjing, China, 2008b.

Tsoukalas, L.H. and Uhrig, R.E., *Fuzzy and Neural Approaches in Engineering.* Wiley, New York, 1997.

Chapter 6

Economics and Energy Transitions

6.1 Coloring Energy Transitions

The global call for a "green energy transition" is often presented as an undeniable imperative: to eliminate emissions, a goal that in practice means the wholesale abandonment of fossil fuels. Unlike historical energy shifts – such as the Promethean mastery of fire or the harnessing of atomic power, both rooted in fundamental physical principles like energy density and entropy production – this contemporary transition is uniquely crafted by policy. Propelled by fears of apocalyptic climate scenarios, vast monetary incentives including subsidies and tax credits, and promises of wealth redistribution from developed to developing nations, it has garnered immense worldwide interest. Yet, this chapter contends that humanity's capacity to achieve this sweeping global imperative by the mid-21st century is severely limited. More critically, it argues that the pursuit of "Net Zero by 2050" could paradoxically entail dangers far exceeding the perceived risks of climate change, a concern underscored by the following points.

First, a sobering look at energy history and population reveals a stark challenge. In 1900, a world of 1.6 billion inhabitants produced negligible global emissions, largely in the absence of widespread fossil fuel consumption. To advocate for the elimination of fossil fuels by 2050 is to imply a return to an energy paradigm that historically supported a fraction of today's global population. The notion of sustaining over eight billion people on an energy footing comparable to that of 1900, without proven, hyper-scalable alternatives, raises the specter of an unprecedented global contraction, not a sustainable transition.

Second, despite significant efforts, the empirical trend in global emissions is moving in the opposite direction of stated goals. In 2004, global CO_2 emissions stood at approximately 28.8 billion tones. Two decades later, in 2024, and after an estimated investment exceeding ten trillion dollars in the green energy transition, these emissions have climbed to 37.4 billion tonnes. This persistent increase, regardless of geographical shifts in emission sources, calls into question the efficacy of current strategies. This chapter will explore the argument that addressing both global energy demand and climate risks may require a more decisive turn towards an accelerated Atomic energy transition, leveraging nuclear power far beyond its currently marginal global contribution.

Since 1900, the world has shown great ingenuity in discovering and using fossil fuels. From the 1850's on and at the early part of the 20th century, coal was king. After World War II, oil became the dominant fuel based on a transnational, transoceanic transport system and a global supply chain with a plethora of products essential for growing economies which are hungry for transport fuels, fertilizers, modern materials and power generation.

In Table 6.1, we juxtapose global population, oil consumption, and total energy consumption in units of barrel of oil equivalent (this is as if all energy needs were obtained from oil). It is easy to see by inspecting the numbers that in 120 years, the global population grew fivefold. The global energy consumption (as given by the daily average for a year in barrels of oil equivalent) grew by more than 20 times and the actual oil consumption by several hundred times. The data clearly shows that hydrocarbon growth led to more than quadrupling of the size of human population. Conversely, any draconian restriction in the use of fossil fuels could spell disaster for humanity.

The small amounts of renewables to date should be of concern to anyone counting on these for the future. For one thing, ramping up new renewables to amounts that can be expected to make a significant contribution is likely to take many years. For another, renewables require fossil fuels for their creation. Solar, wind, and batteries are very much tied to the Promethean energy transition. They require a lot of fossil fuels. Only the location of their use may be

Table 6.1. World population, oil consumption, energy consumption in oil-equivalent*.

Year	World population (Billion)	Oil consumption per day (Million BBL/D)	Energy consumption in oil-equivalent per day (Million BOE/D)
1900	1.6	0.2	12.5
1920	1.8	2.0	25.0
1940	2.3	4.0	41.0
1960	3.0	20.0	64.0
1980	4.4	63.0	127.0
2000	6.1	76.0	191.0
2020	7.9	88.0	250.0
2024	8.1	102	289.0

Note: *Approximate numbers derived from the databases of the Energy Information Administration (EIA) of the US Department of Energy and data from BP Statistical Review.

altered, not their use. In other words, we may be just selecting using coal, oil, and gas in China and the Global South.

Modernity is based on cheap energy, especially cheap petroleum, and without increasing amounts of it year after year, cracks may appear in the edifice of modern economies. The AI-Energy nexus could offer means to contain these cracks through effective demand-side management or better management of the Atomic energy transition. As seen in Figure 6.1, the per capita use of energy has been increasing since the early 1900s with oil and coal being protagonists and natural gas following in recent years.

The period of rapid energy growth following World War II, seen in Figure 6.1, is likely to be repeated in a more intense form in the next several decades due to AI energy demands. The rapid post 1945 energy growth allowed much manual work to be performed by machines for the reconstruction of war-torn areas. Thus, there appeared to be considerable growth in human efficiency, but such growth is not likely to succeed without moving into the Atomic energy transition, even with lower rates of Global GDP growth.

Even the period between 1980 and 2000 may be misleading for predicting future patterns because this period occurred before the

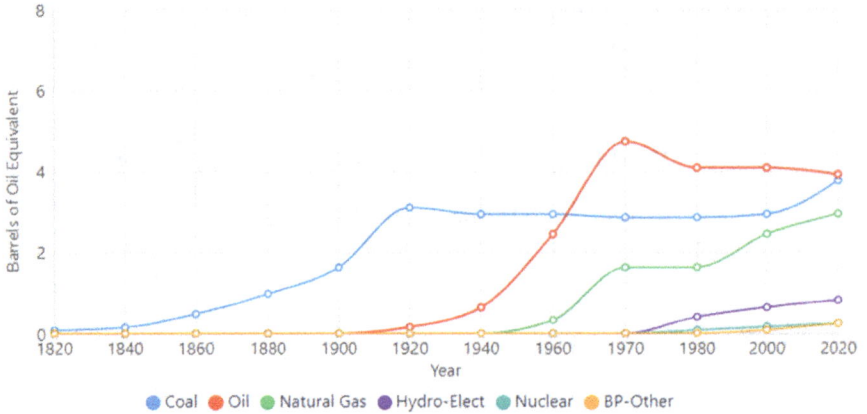

Figure 6.1. Per capita use of coal (blue), oil (red), natural gas (green), hydro (purple), nuclear (lowest green curve) and biofuels (yellow).

huge increase in international trade that followed the collapse of the Soviet Union. Once international trade with less developed nations increases, we can expect these nations will want to increase their energy consumption in any way that is possible, including using more coal.

Energy policy is typically formulated and implemented within national frameworks of responsibility. Multilateral organizations such as the European Union, the OECD, or the BRICS may provide directives, technical support, and investment guidance, hence streamline their energy policy objectives and aligning goals to support additional constraints. For instance, a trillion-dollar EU investment in wind and solar, incentivized many EU member states to turn to wind and solar to bring home part of that capital. But individual countries behave very differently. While some countries may continue to grow using coal, other countries will flounder when hit by high oil and natural gas prices. Some countries may encounter major difficulties in the years ahead, even though they have so far been untouched. The precarious debt situations of several countries leave them vulnerable to disruptions.

Another false inference might be that per capita oil consumption has declined in the past (see Figure 6.1) so future declines should not be a problem. If enough energy assets including suppliers, generators transmission capacities, and network links to the information infrastructure, sellers can be price-takers and consumers can enjoy affordable energy. Growing connectivity is important so that there is a functional and effective market structure for buyers, sellers, brokers, regulators, and other service entities. A conjectural look in the future of energy identifies two possible solutions forward that can work synergistically:

1. Fully embrace the Atomic Energy Transition (including its capacity to generate new fuel) while responsibly grow Promethean Energy as needed and as is feasible.
2. Use advanced markets structures and AI for rational allocation of energy.

Example 6.1. Population Growth Since 1900.

Using Table 6.1 develop a model that best fits the population data and predicts future populations. What does the model predict for the world population in 2030, 2040, and 2050?

Solution: This dataset suggests an increasing trend in world population, so let's find the best-fit model. A common approach is to apply exponential or logistic growth models, but polynomial regression could also provide insight.

For an exponential fit, we assume:

$$P(t) = P_0 \, e^{rt} \qquad (6.1.1)$$

where

P_0 is the initial population,
r is the growth rate, and,
t is time (year).

Taking the natural logarithm of both sides of Eq. (6.1.1), we have

$$\ln P = \ln P_0 + rt \qquad (6.1.2)$$

This turns our problem into a linear regression, where we fit $\ln P$ against t.

To compute the regression parameters, we rewrite equation (6.1.2) as

$$y = a + bt$$

where $y = \ln P$, $a = \ln P_0$ and $b = r$.

Using least squares, we compute a and b by solving:

$$b = \frac{\sum_{i=1}^{8} (t_i - \bar{t})(y_i - \bar{y})}{\sum_{i=1}^{8} (t_i - \bar{t})^2} \tag{6.1.3}$$

$$a = \bar{y} - b\bar{t} \tag{6.1.4}$$

where, $\bar{t}$ and $\bar{y}$ are the means of the t-values and y-values computed as,

$$\bar{t} = \frac{1900 + 1920 + 1940 + 1960 + 1980 + 2000 + 2020 + 2024}{8} = 1968$$

Using, the values of ln (P) for the populations given in Table 6.1, i.e.,

$$\ln(1.6) = 0.470, \quad \ln(1.8) = 0.588, \quad \ln(2.3) = 0.832,$$

$$\ln(3.9) = 1.099, \quad \ln(44) = 1.481, \quad \ln(6.1) = 1.808,$$

$$\ln(7.9) = 2.068, \quad \ln(8.1) = 2.092$$

we obtain

$$\bar{y} = \frac{0.470 + 0.588 + 0.832 + 1.099 + 1.481 + 1.808 + 2.068 + 2.092}{8}$$

$$= 1.305$$

Thus, from equation (6.1.3), we obtain

$$b \approx 0.0081$$

while, from equation (6.1.4), we have

$$a = \bar{y} - b\bar{t}$$

$$a = 1.305 - (0.0081 \times 1968)$$

or

$$a \approx -14.6$$

Since $a = \ln P_0$, this implies that $-14.6 = \ln P_0$ and thus exponentiating both sides we obtain,

$$P_0 = e^{-14.6}$$

Therefore, the exponential model of global population growth over the last nearly century and a quarter is

$$P(t) = e^{-14.6} \cdot e^{0.0081t} \qquad (6.1.5)$$

with $P_0 = e^{-14.6}$.

Figure 6.2 is a plot of world population from 1900 on. The model of equation (6.1.5) suggests continued exponential growth of the global population, though in reality, factors like fossil fuel limitations and other resource limits as well as policy changes might slow the rate.

For $t = 2030$, 2040 and 2050 the global population is obtained by plugging into (6.1.5)

$$P(2030) = e^{-14.6}e^{0.0081(2030)} \approx 8.94 \text{ billion}$$

$$P(2040) = e^{-14.6}e^{0.0081(2040)} \approx 10.00 \text{ billion}$$

$$P(2030) = e^{-14.6}e^{0.0081(2050)} \approx 11.20 \text{ billion}$$

Alternatively, a logistic model would be more realistic if population growth is slowing down, that is,

$$P(t) = \frac{K}{1 + A\,e^{-rt}}$$

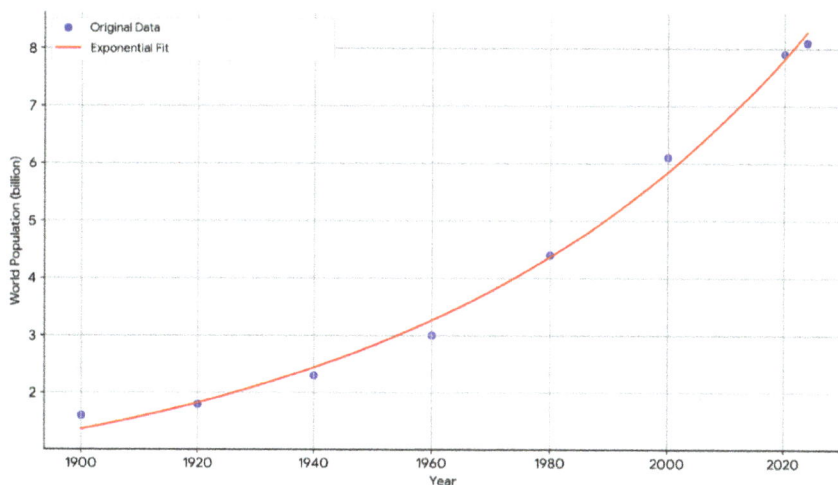

Figure 6.2. World population in exponential growth since the early part of the 20th century.

where K represents the carrying capacity (see problems at the end of Chapter 6).

Example 6.2. Oil Consumption Per Day since 1900.

Using the data of Table 6.1, what is a good fitting model for global oil consumption per day since 1900?

Solution: Again, since we have major growth (after about 1920), an exponential model may be a best fit. Let $B(t)$ stand for the barrels of oil per day (at year t), then the model would be

$$B(t) = B_0 e^{rt} \qquad (6.2.1)$$

where

B_0 is the initial oil consumption per day at year 1900,
r is the growth rate, and,
t is time.

Following the steps outlined in Example 5.1, we compute the regression parameters first by taking the natural log of both sides of Eq. (6.2.1),

$$\ln B = \ln B_0 + rt \qquad (6.2.2)$$

We write Eq. (6.2.2) as

$$y = a + bt$$

where $y = \ln B$, $a = \ln B_0$ and $b = r$.

Using least squares, we compute a and b by solving:

$$b = \frac{\sum_{i=1}^{8}(t_i - \bar{t})(y_i - \bar{y})}{\sum_{i=1}^{8}(t_i - \bar{t})^2} \qquad (6.2.3)$$

$$a = \bar{y} - b\bar{t} \qquad (6.2.4)$$

Where, $\bar{t}$ and $\bar{y}$ are the means of the t-values and y-values computed as,

$$\bar{t} = \frac{1900 + 1920 + 1940 + 1960 + 1980 + 2000 + 2020 + 2024}{8} = 1968$$

Using, the values of ln (B) for the bbl given in Table 6.1, i.e.,

$$\ln(0.2) = -1.609, \quad \ln(2.0) = 0.693, \quad \ln(4) = 1.386,$$
$$\ln(20) = 2.996, \quad \ln(63) = 4.143, \quad \ln(76) = 4.330,$$
$$\ln(88) = 4.477, \quad \ln(102) = 4.625$$

we obtain

$$\bar{y} = \frac{-1.609 + 0.693 + 1.386 + 2.996 + 4.143 + 4.330 + 4.477 + 4.625}{8}$$
$$= 2.630$$

Thus, from Eq. (6.2.3), we obtain,

$$b \approx 0.0469$$

while, from Eq. (6.2.4), we have

$$a = \bar{y} - b\bar{t}$$
$$a = 2.630 - (0.0469 \times 1968)$$

or,

$$a \approx -89.66$$

Since $a = \ln B_0$, this implies that $-89.66 = \ln B_0$ and thus exponentiating both sides we obtain,

$$B_0 = e^{-89.66}$$

Therefore, the exponential for oil consumption is:

$$B(t) = e^{-89.66} e^{0.0469t} \tag{6.2.5}$$

with, $B_0 = e^{-89.66}$

Figure 6.3 shows the exponential nature of oil consumption growth in the 20th century. The R-squared value for this fit is 0.9125, which indicates a strong fit to the provided data. It is important to note the similarity of oil consumption exponential with the population curve of Figure 6.2. In a simplified sense, they both follow a similar trajectory. On the basis of this model, *ceteris paribus*, the oil

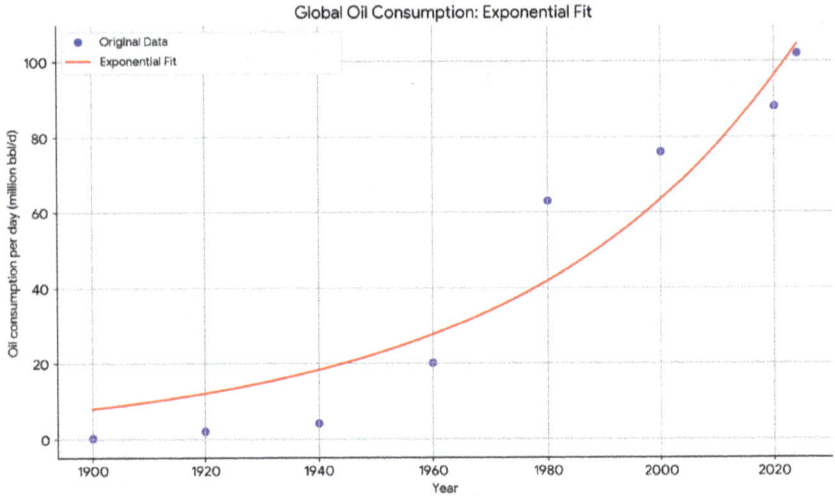

Figure 6.3. World oil consumption in exponential growth since the early part of the 20th century.

consumption in 2030, 2040 and 2050 is predicted to be (in million barrels per day, bbl/day):

2030: 118.15 million bbl/day

2040: 145.47 million bbl/day

2050: 179.11 million bbl/day

Given the projected future demand, it is highly doubtful that the global oil supply can adequately meet this need (King and Murray, 2012). Consequently, diversifying away from oil is not merely important, but absolutely crucial. Natural gas emerges as a viable transition fuel for the next quarter-century, bridging the gap until the full deployment of the Atomic energy transition.

Example 6.3. Per capita oil consumption per day, and per capita energy consumption per day.

    Using Table 6.1 and the population data from the model of example 6.1 estimate past and future (2030, 2040, 2050) for

(a) Per capita annual oil consumption in bbl/day,
(b) Per capita annual energy consumption in oil equivalent (boe/ day) terms.

Solution: From Example 6.1 the population model is

$$P(t) = e^{-14.6} \cdot e^{0.0081t}$$

which simplifies to,

$$P(t) = e^{0.0081t - 14.6}$$

We evaluate $P(t)$ for $t = 2030$, 2040 and 2050:

$$P(2030) = e^{0.0081 \cdot 2030 - 14.6} \approx 8.94 \text{ billion}$$

$$P(2040) = e^{0.0081 \cdot 2040 - 14.6} \approx 10.00 \text{ billion}$$

$$P(2050) = e^{0.0081 \cdot 2050 - 14.6} \approx 11.19 \text{ billion}$$

We may use exponential fitting following the treatments in the previous examples, that is,

$$C(t) = C_0 \cdot e^{k(t-1900)}$$

where

$C(t)$ is the oil or energy consumption at time t,
C_0 is the consumption at year 1900,
k is the exponential growth rate and t is the year.

For the oil and energy consumption data, it is best to use the data points from 1960 to 2024, for better prediction accuracy, as earlier data differs greatly in scale. Thus, we have

$$C(t) = C_0 \cdot e^{k(t-1960)}$$

(a) Oil Consumption

If we use the year 1960, where the oil consumption was 20 bbl/day and the year 2024 where the oil consumption was 102 bbl/day, we get

$$\frac{102}{20} = e^{k(64)} \Rightarrow 5.1 = e^{k(64)} \Rightarrow k \approx \frac{\ln(5.1)}{64} \approx 0.026$$

$$C_0 = 20$$

So,

$$C_{\text{oil}}(t) = 20 \cdot e^{0.026(t-1960)}$$

(b) Oil-equivalent energy consumption

If we use the year 1960, where the oil-equivalent energy consumption was 64 boe/day and the year 2024 the oil-equivalent energy consumption was 289 boe/day, we get

$$\frac{289}{64} = e^{k(64)} \Rightarrow 4.52 = e^{k(64)} \Rightarrow k \approx \frac{\ln(4.52)}{64} \approx 0.0247$$

$$C_0 = 64$$

So,

$$C_{\text{energy}}(t) = 64 \cdot e^{0.0247(t-1960)}$$

To compute per capita oil and energy consumption:

$$\text{Per capita consumption} = \frac{C(t)}{P(t)}$$

- Oil consumption for 2030:

$$C = 20 \cdot e^{0.026(2030-1960)} = 20 \cdot e^{1.82} \approx 123.44$$

$$\text{Population} \approx 8.94 \text{ billion}$$

$$\text{Per capita: } \frac{123.44 \times 10^6}{8.94 \times 10^9} \approx 0.0138 \frac{\text{bbl}}{\text{day}}/\text{person}$$

- Oil consumption for 2040:

$$C = 20 \cdot e^{0.026(2040-1960)} = 20 \cdot e^{2.08} \approx 160.09$$

$$\text{Population} \approx 10.00 \text{ billion}$$

$$\text{Per capita: } \frac{160.09 \times 10^6}{10.00 \times 10^9} \approx 0.0160 \frac{\text{bbl}}{\text{day}}/\text{person}$$

- Oil consumption for 2050:

$$C = 20 \cdot e^{0.026(2050-1960)} = 20 \cdot e^{2.34} \approx 207.62$$

$$\text{Population} \approx 11.19 \text{ billion}$$

$$\text{Per capita: } \frac{207.62 \times 10^6}{11.19 \times 10^9} \approx 0.0186 \frac{\text{bbl}}{\text{day}}/\text{person}$$

- Energy consumption for 2030:

$$C = 64 \cdot e^{0.0247(2030-1960)} = 64 \cdot e^{1.729} \approx 360.64$$

$$\text{Population} \approx 8.94 \text{ billion}$$

$$\text{Per capita: } \frac{360.64 \times 10^6}{8.94 \times 10^9} \approx 0.0403 \frac{\text{boe}}{\text{day}}/\text{person}$$

- Energy consumption for 2040:

$$C = 64 \cdot e^{0.0247(2040-1960)} = 64 \cdot e^{1.976} \approx 461.69$$

$$\text{Population} \approx 10.00 \text{ billion}$$

$$\text{Per capita: } \frac{461.69 \times 10^6}{10.00 \times 10^9} \approx 0.0462 \frac{\text{boe}}{\text{day}}/\text{person}$$

- Energy consumption for 2050:

$$C = 64 \cdot e^{0.0247(2050-1960)} = 64 \cdot e^{2.223} \approx 591.04$$

$$\text{Population} \approx 11.19 \text{ billion}$$

$$\text{Per capita: } \frac{591.04 \times 10^6}{11.19 \times 10^9} \approx 0.0528 \frac{\text{boe}}{\text{day}}/\text{person}$$

Year	Global Per Capita Oil Consumption (bbl/day)	Global Per Capita Energy Consumption (boe/day)
2030	0.0138	0.0403
2040	0.0160	0.0462
2050	0.0186	0.0528

The annual per capita numbers are obtained by multiplying the two columns above by 365 days/year, obtaining the following table.

Year	Global Per Capita Oil Consumption (bbl/year)	Global Per Capita Energy Consumption (boe/year)
2030	5.0	14.7
2040	5.8	16.9
2050	6.8	19.2

It should be noted that in the United States the average per capita oil consumption is approximately 22 bbl/y/per person, while Japan, S. Korea and EU consume approximately half of the US average (around 10 bbl/y/per person).

6.2   Metrics

The enduring insights of the great Austrian economist Joseph A. Schumpeter highlight the central role of *innovation* in shaping modern economic systems. He posited that capitalism is fundamentally driven by a "perennial gale of creative destruction," a dynamic process where groundbreaking innovations revolutionize the economic structure from within. This process simultaneously renders old industries, firms, and technologies obsolete while creating entirely new ones. Though often disruptive, leading to short-term pains like job losses or business failures, this continuous cycle ultimately propels long-term economic growth, enhances productivity, and elevates living standards by reallocating crucial resources – capital, labor, materials, and energy – to more productive uses. It is precisely within this framework of Schumpeterian disruption that the AI-Energy nexus must be understood if we are to grasp its profound call for innovation as the path forward.

Indeed, energy transitions represent a physical manifestation of this evolutionary economic process. Drawing insights from thermodynamics, particularly the concept of *entropy production*, clarifies that all economic activity, including Schumpeter's relentless cycle of creative destruction, is an *irreversible process*. It inherently consumes available energy and materials, contributing to the overall increase of entropy. The shift towards fully harnessing Atomic energy, including its capacity to breed new fuel, represents a pivotal step in this ongoing, dynamic transition, offering the physical foundation for the next stage of economic development and human efficiency.

Just as living organisms maintain a low-entropy state locally by increasing the entropy of their surroundings (by consuming high-quality energy and expelling waste), economic systems driven by creative destruction create pockets of "order" (new products, efficient processes, wealth) by accelerating the overall entropy production of the larger system (the planet). Innovation can be seen as a temporary, local reduction in economic "disorder" or inefficiency, but this often comes at the cost of increased global entropy production through resource extraction and waste generation (Schumpeter, 1964).

Several metrics are available for comparing technologies and fuels used in energy production. One of the most common is the Levelized Cost of Energy (LCOE). It is a comprehensive metric that estimates the average total cost of building and operating a power plant over its lifetime, divided by the total electricity output. For instance, if a Small Modular Reactor produces 250 million MWh over its operating life estimated at 100 years, the Levelized Cost of Energy for the SMR to a first approximation is:

$$\text{LCOE} = \frac{\$5 \text{ bilion}}{250{,}000{,}000 \text{ MWh}} = \frac{\$40}{\text{MWh}}$$

In reality, though, this approach is too simple; it may not capture adequately the various expenses involved in building and operating a plant generation plant. Additional limitations have to do with uneven types of comparisons. For instance, intermittent sources like wind and solar, require dispatchable sources like gas turbines or hydro of grid batteries to operate when the sun does not shine and the wind does not blow. *LCOE* focuses primarily on the cost of producing energy at the "busbar" that is, at the plant gate, without fully considering the value of that particular energy unit to the grid edge.

A more accurate definition of the *Levelized Cost of Energy* (LCOE) is designed to represent the average total cost of building and operating a power plant over its lifetime. The sum of all costs is levelized per unit of electricity produced, taking into account the time value of money (discount rate). The most commonly used formula is

$$\text{LCOE} \equiv \frac{\sum_{t=1}^{n}\left[\frac{I_t+M_t+F_t+D_t}{(1+r)^t}\right]}{\sum_{t=1}^{n}\left[\frac{E_t}{(1+r)^t}\right]} \tag{6.1}$$

where

LCOE $\equiv$ Levelized Cost of Energy (usually in \$/MWh or \$/kWh)
$I_t \equiv$ Investment expenditures (capital costs) in year t
$M_t \equiv$ Operations and maintenance (O&M) expenditures in year t
$F_t \equiv$ Fuel expenditures in year t (if applicable)
$D_t \equiv$ Decommissioning costs
$E_t \equiv$ Electricity generation (energy produced) in year t (in MWh or kWh)
$r \equiv$ Discount rate or Weighted Average Cost of Capital
$n \equiv$ Economic lifetime of the system (in years)

The numerator in Eq. (6.1) that is, $\sum_{t=1}^{n}\left[\frac{I_t+M_t+F_t}{(1+r)^t}\right]$, represents the present value of the total costs. It adds all costs, $I_t + M_t + F_t$, incurred over the lifetime of the project ($t = n$) and discounts them back to their present value by dividing with the $(1 + r)^t$ term using the discount rate, r. This is tantamount to the assumption that the costs incurred further in the future are worth less in today's money. On the other hand, investment expenditures capture the upfront capital costs for planning, permitting, construction, equipment, and initial commissioning. For simplicity, sometimes, the entire capital cost is assumed to occur in the first year of the project, that is $t = 0$. Operations and Maintenance (O&M) indicated by the term, M_t, include fixed costs regardless of output, e.g., salaries, property, taxes, as well as variable O&M (costs related to output, e.g., consumables, wear-and-tear); while fuel costs, F_t, is the cost of the fuel used to produce electricity. It should be noted that coal, natural gas, oil, biomass or uranium may be used to produce renewables, but these sources once bought and installed have zero fuel costs. If we exclude the costs of the fuels used to manufacture them, then the LCOE for renewables may look relatively low. On average, fuel costs for coal, gas and oil significantly exceeds about 30% of total costs in fossil fuel technologies for generating electricity. Decommissioning costs D_t may be included as a future operating cost M_t at the end of the plant's life, also discounted to present value, but this is a practice that again, may disguise some of the real costs involved in wind and solar.

The denominator of Eq. (6.2) represents the sum of all electrical energy produced each year E_t over the lifetime of the generation unit and, also, discounts it back to its present value. This may account for the idea that a unit of energy produced earlier in the life of the generator may be considered more "valuable" that a unit of energy produced far in the future due to the time value of money. The electrical energy generated in each year depends on the plant's Capacity (MWe) but also on the Capacity Factor, that is the percentage of time in a year that the plant is outputting energy at its nominal value. This is mathematically stated as

$$E_t = [\text{Capacity (MWe)}] \times [\text{Capacity Factor (\%)}] \times \left[8760 \frac{\text{hours}}{\text{year}} \right]$$

An additional factor to consider is the discount rate, r. This is a crucial factor which reflects the cost of capital (how much it costs to borrow or use equity) and the investor's required rate of return, often represented as the Weighted Average Cost of Capital (WACC). A higher discount rate increases the *LCOE*, especially for capital-intensive projects like nuclear units with long lifetimes, as it makes future revenues less valuable compared to upfront costs.

For nuclear power generation, the fuel cost is paid upfront. When a nuclear reactor is built, the original fuel should cover its fuel needs for 20 years. Hence, the variability of fuel costs is eliminated and a different model for LCOE is used, typically using a Fixed Charge Rate (FCR), assuming constant annual costs and generation. FCR bundles together various financial and tax-related factors. The components of FCR can vary depending on the specific financial model, tax laws, depreciation schedules and desired return rates. In more detailed considerations, FCR needs to include the Capital Recovery Factor (CRF) which converts an initial investment into a series of equal annual payments over a given period. This is analogous to paying a loan. The CFR is defined as:

$$\text{CRF} \equiv \frac{r(1+r)^n}{(1+r)^n - 1} \tag{6.2}$$

where

$r \equiv$ Discount rate (or Weighted Average Cost of Capital, WACC)
$r \equiv$ Project economic lifetime (in years)

In the WACC factor, we often embed the Return on Equity (ROE) and Cost of Debt (IR), which is the discount rate, r, used in the CRF (Eq. (6.2)). The WACC terms is thus defined as

$$\text{WACC} \equiv \left[\frac{E}{V}\right] \times R_e + \left[\frac{D}{V}\right] \times R_d \times (1 - T_c) \qquad (6.3)$$

where

$E \equiv$ Market value of equity
$D \equiv$ Market value of debt
$V \equiv E + D$ Total market value of financing
$R_e \equiv$ Cost of equity
$R_d \equiv$ Cost of debt
$T_c \equiv$ Corporate tax rate

Capital investments can be depreciated for tax purposes, creating a "tax shield" b reducing the taxable income; which effectively reduces the overall cost of capital. The fixed charge rate, FCR, must account for the present value of these depreciation benefits. To account for these effects, we used the Present Value of Depreciation (PVD) divided by the Total Capital Cost (TCC) of a project which then can be factored into depreciation schedules. In addition, central (federal), state and local taxes affect the after-tax cost of debt and the require return on equity as well as property taxes and insurance costs. A more elaborate formula for FCR is given as follows.

$$\text{FCR} \equiv \left(\frac{\text{CRF}}{1 - T_R \times \text{PVD}}\right) \times \left(\frac{1}{1 - T_R}\right) \times (1 + \text{CFF})$$
$$+ (\text{Prop. Tax Rate}) + (\text{Insur. Rate}) \qquad (6.4)$$

where

CRF $\equiv$ Capital Recovery Factor
$T_R \equiv$ Effective income tax rate (considering federal and state)
PVD $\equiv$ Present Value of Depreciation (per dollar of capital cost, reflecting tax savings from depreciation)

CFF $\equiv$ Construction Finance Factor (accounts for interest during construction, effectively increasing the "all-in" capital cost)

The complexity of Eq. (6.4) in defining FCR arises from the aim to capture a variety of disparate costs (one-time capital, ongoing debt payments, equity returns, taxes, depreciation benefits) into a single, levelized annual percentage of the initial investment. Financial modeling software and proprietary tools used by utilities are typically used to calculate FCR accurately, as it requires iterating through cash flows over the project's lifetime and applying specific tax and financing rules. This is a highly aggregated factor that reflects the "cost of capital" (debt and equity) plus the various fixed annual charges that are directly proportional to or based on the initial investment, all expressed as a percentage of that initial investment (NREL, 2025).

6.3 Allocation of Energy Through Market Structures

The AI-energy nexus pivots on market structures to enable a more efficient allocation of energy, particularly electricity, by influencing pricing, competition, investment, and consumer behavior. Competition incentivizes generators to produce electricity at the lowest possible cost (productive efficiency). It drives innovation in generation technologies (dynamic efficiency) and can lead to more diverse and flexible energy auctions which in turn produce pricing signals that can be used in multitude of ways to minimize entropy production. In competitive markets, prices are determined by supply and demand. These price signals are crucial for efficient allocation. For instance, high prices during peak demand signal generators to increase output and consumers to reduce consumption. Low prices during off-peak or high renewable output periods encourage consumption (e.g., EV charging) or storage. Sustained high prices in a region signal investors to build new generation capacity, while low prices might discourage investment in less efficient plants.

Well-designed market structures, particularly electricity markets, move beyond simple bilateral agreements to create competitive environments. These structures use transparent price signals and specific market mechanisms to incentivize efficient generation, optimal resource allocation, necessary investment in grid infrastructure and flexible resources, and active participation from consumers and distributed energy assets. This leads to a more agile, resilient, and ultimately more efficient energy system.

How much to charge for a unit of energy? What is a fair price for, say, a kilowatt-hour or a barrel of oil or a thousand cubic meters of natural gas? Should there be a single price or multiple prices? What is the elasticity of demand as a function of price? How is price discovered?

We turn to economics to answer such questions and to energy economics in particular, since the economics of energy and electricity have unique features and characteristics. Energy investments have long horizons. The design and operating time scales of modern nuclear power plants, for example, may span several decades and approach, if not exceed, one hundred years. Although energy investments offer good returns and produce much needed innovation and employment, energy's unique long-term scales and thermodynamic considerations call for a focus on energy decisions that pertain to the short run.

In smart energy economics a variety of assumptions are made and should be clearly understood. For example, in competitive electricity markets assumptions about assets in place are critical for the functioning of the market. If a sufficient number of physical assets such as generators and transmission capacities are in place the providers of these assets can be price-takers. If this is not the case, they can become price-makers thus diminishing the competitive nature of the market. Power plants, the transmission grid, substations, distribution lines should be there and ready to be used. Assumptions about connectivity are also there so that a market structure for buyers, sellers, brokers, regulators and other service entities is clear and complete.

Generating, transmitting and distributing electricity through the grid is undergoing a transition with pricing becoming more time-varying, and dynamic. Time-varying prices are expected to incentivize new generation and grid technologies and direct power usage away from peak demand while, at least in the long run, deliver lower energy costs to consumers.

Since the early days of electricity, power utilities were charged with producing electricity as a commodity that is, as a mass-produced, unspecialized product with smaller but stable profit margins, and little importance for factors such as marketing and brand name. Utilities bundled all functions necessary to achieve production, transmission and distribution of this commodity in a structure called *vertically integrated monopoly*. The price to the consumers was determined by dividing the generated kilowatt-hours over a period of time with all associated costs plus profit on an average sense. This was more or less how price-discovery worked to produce a unit price that was charged to the consumers.

In the deregulated or liberalized electricity markets, price-discovery assumes new meaning. It is achieved through unbundling functions and establishing competitive market structures with a large number of participants bidding for energy and related services as well as independent system operators clearing contracts and overseeing the process. Marginal rather than average cost is the standard determinant of competitive pricing.

In this chapter, we examine aspects of the economics of smart energy with emphasis on price discovery. Two different models of electricity markets are examined together with their associate price discovery mechanisms: utilities functioning as regulated monopolies and utilities functioning as deregulated energy systems. It should be noted that much of what is discussed for electricity is also relevant for other energy markets, particularly natural gas.

How is price determined when electricity is sold in markets that function as regulated monopolies? Be they investor-owned, cooperatives, or publicly-owned companies, regulated utilities are the sole supplier of electricity within a geographic area and price is largely based on average costs plus profit.

Electric power systems are peak-demand constrained systems where peak demand cannot exceed the total power generated at any given time. To reduce generation-demand mismatch, regulated utilities devise special rate structures offering discounted prices during certain periods of day or night to incentivize power consumption while higher prices at other times incentivize reduced consumption thus reducing the risk of system collapse due to the inability to meet demand.

In a broad sense, regulated utilities determine the cost of a unit of electric energy, say a kilowatt-hour, by aggregating all costs plus profit and dividing the total amount by the number of produced kilowatt-hours over a period of time (usually one or more years). Utility customers are categorized into different groups or "sectors" such as residential, commercial or industrial sectors. A price structure is established with different rates charged for different sectors. Further distinctions are drawn for each sector and the price may vary for different intervals over a 24-hour period.

What determines total cost and what are the individual constituent costs determining total cost? In general, there are three different categories of costs: capital costs, financing costs, and operating costs.

Capital costs include the cost of engineering, procurement and construction, the cost of land, buildings and the cooling and transmission infrastructures. Cost escalation due to rise in material and labor expenditures and inflation as well as decommissioning, environmental reclamation and waste disposal costs can be included in this category.

Financing costs include the interest charged on money borrowed, debt-to equity ratios, capital cost recovery, risk-based financial fees and cost recovery. Financing and capital costs together are often referred to as *project costs*.

Operating costs are associated with the operation and maintenance of the system and include, but are not limited to, fuel and labor cost, maintenance and local taxes. When comparisons of different types of power plants are made, the concept of *levelized costs* is used. This concept refers to the average costs of producing electricity over a plant's lifetime.

Regulated utilities are vertically integrated, that is, they have all functions bundled together in a unified framework. During normal operations, a decision needs to be made as to which generating unit should be connected to the grid. When several alternatives are present, a method called *economic dispatch* is used in decision-making. Control engineers, in what is usually called the "dispatch center" of the utility, are charged with decision making authority.

Economic dispatch uses forecasted demand for each hour (or half hour) in the next 24-hour period to schedule generating units (for tomorrow) and makes decisions about dispatching generating units in the current 24-hour period (operating day). The decision to dispatch is based on algorithms and heuristics that seek optimal or near optimal solutions that prioritize the available generating units through a number of criteria that include, but are not limited to, the cost of fuel and operation and the location of the generator with respect to the transmission network. Economic dispatch shifts generation from higher cost to lower cost units with overall cost minimization goals. Best practices from economic dispatch provide the foundation for the development of pool dispatch and pool-based competitive electricity markets.

At any interval time, power plants have a marginal cost of generating power. Aggregating generator costs produces the generation "merit order," from least to most expensive. This merit order defines the short-run marginal-cost curve which determines the supply of power. Power pools which are associations of two or more power systems having agreements to coordinate operations provide the model for achieving the most efficient dispatch given the short-run marginal costs of power supply. By extension, in deregulated systems, the independent system operator controls operation of the system to achieve an optimal match between supply and demand based on bids from market participants whose transactions essentially match schedules with any associated bids. In these systems, the economics are determined by market structures of higher complexity but in principle similar to economic dispatch, at least in the short run. Information is available to the dispatcher via bids and these are used to determine which plants will run at any given interval of time. For the same load, the least-cost dispatch of regulated utilities and competitive electricity market dispatch ought to be the same.

6.4   Energy Markets Deregulated Systems

The economics of competitive electricity markets emphasize functional unbundling and an information architecture featuring financial instruments to bridge the physical pillars of generation, transmission and distribution. An independent system operator with coordination, oversight and dispatch authority is essential in this architecture as well as a short-run market of bids and contracts which serves as the foundation for long-run contracts as well as markets for transmission and congestion control. Secondary markets to achieve rational allocation of investment capital by addressing longer term risks and anticipated volatilities intertemporally can use, but do not have to, this architecture.

In the 1990s successful market design frameworks for electricity services were proposed by Hogan and others (Hogan, 1998). Such approaches informed legislative and implementation initiatives for competitive electricity markets in a number of countries, states and regions. In effect, market structures supporting bid-based wholesale transactions for electric energy were viewed as an extension of economic dispatch. All approaches require an independent system operator (ISO) to schedule generation dispatch and clear transactions over the grid, while maintaining reliability and quality of service. Based on short-term economic analysis, the opportunity cost is assessed and upon it the longer-term market structures are built.

Similar to the deregulation of the railroad industry where different train companies offering competitive services use a common track infrastructure independently managed, electric utility deregulation introduces competition in generation and retail services. The transmission and distribution infrastructure is under the secure purview of an independent system operator.

Competitive electricity markets share a number of structural features including a short-term spot market operating essentially as an extension of the traditional economic dispatch. Spot markets offering transactions related to losses and congestion are part of competitive markets as well as settlement systems using day-ahead bidding, pricing and contracts, with real-time balancing at real-time market prices.

Following Hogan (1998), Figure 6.4 illustrates the structure of a competitive electricity market, where generating companies (Genco) are distinguished from transmission grid companies (Gridco) and distribution companies (Disco) which include different marketers and retailers. At the bottom, we find individual customers (Cust) of various sizes and characteristics. This structure can readily accommodate charges for loads to recover other unbundled ancillary service costs. In the short-term, the least-cost dispatch and the competitive market dispatch should be the same.

In competitive electricity markets, market participants (generators, marketers, and customers) should be price takers, that is, none should have undue influence on the price of electricity. Over a predetermined period of time (usually half hour or less), the market delivers energy from generators to customers while the independent operator coordinates all bidding and settlement activities and ensures

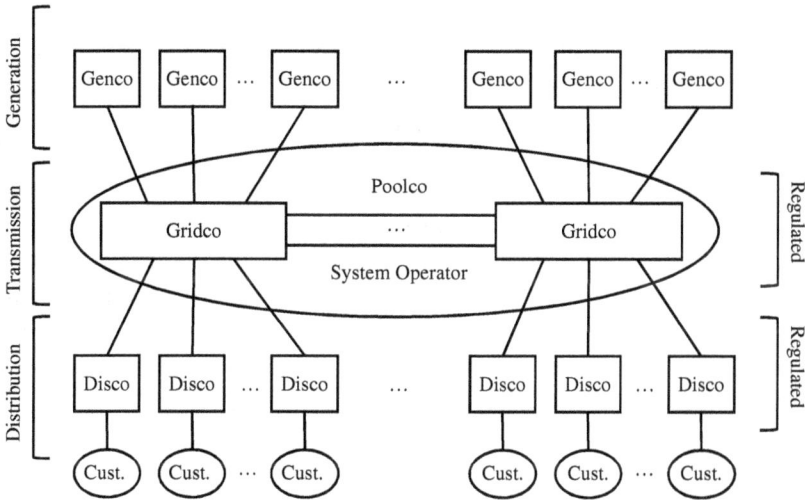

Figure 6.4. Competitive wholesale electricity market structure.

reliable operation. Market participants, through bilateral transactions, provide schedules with associated bids to the independent system operator which are used in a decision process that determines which plants will run at any given interval of time (for example, half hour). The operator achieves efficient match of supply and demand based on the preferences of the participants as expressed in the bids. Unlike traditional economic dispatch, however, in a competitive electricity market marginal-cost information is provided in the form of bids for the right to generate electricity. Each bid defines the minimum acceptable price that the generator would accept to generate electricity. In addition to generation bids, the operator uses transmission congestion contracts to ensure the physical integrity of the system. For power delivery, the laws of nature have to be respected. Ohm's law and Kirchhoff's laws define physical limits to the capacity of the transmission network to deliver power.

The relation between dispatch and transmission is of special interest to the independent system operator. Settlements for generation bids must not threaten the grid. The operator guides dispatch decisions based on generation bids and congestion bids to ensure that the transmission network does not fail because of electrical overloading, violation of frequency and voltage limits, loop currents, and anticipated contingencies that may threaten the grid.

If the system has a sufficient number of participants and market rules (prohibiting collusion) are respected, no generator can determine the marginal plant. Thus, the optimal bid for each generator is the true marginal cost offering buyers and sellers a way of settling at the market equilibrium price.

Since producers and consumers are connected through the grid, where the laws of physics dictate power losses proportional to the square of the length of a conducting line, the marginal cost of delivering power to different locations differs when taking into account such losses and adjusting market equilibrium prices accordingly. In addition, long-distance or location specific constraints may result in higher marginal costs for generators at certain locations. Ideally, power generated at low-cost locations has precedence over high-cost locations but, if in periods of high demand not all the power that could be generated in the low-cost region can be used, the marginal cost in the two locations differs because of congestion. Thus, power from two different locations can have different marginal costs due to congestion-induced costs and a congestion rental or fee can be calculated accordingly (Hogan, 1998).

Through this bidding and settlement process during each interval of time or market period, electricity can be treated as a commodity. In the short run customers and producers pay or receive the true opportunity cost. As supply and demand conditions vary, on a real time basis, there will be fluctuations in the price of electricity and spot market volatility may be the norm, presenting risks for both consumers and producers.

Such risks can be mitigated by long-term and secondary market mechanisms. In theory, long-term contracts and secondary market mechanisms focus on price volatility and risk and provide a price hedge not by managing the flow of power but by managing the flow of information analogous to the manner discussed in the previous chapter.

Thus, a financial/informational perspective of the system emerges with perfectly tradable rights as shown in Figure 6.5. The financial perspective can be extended to pricing signals directing demand automatically.

Price discovery for energy in competitive electricity markets is achieved through complex informational approaches which in effect segregate the physical infrastructure from the financial/informational

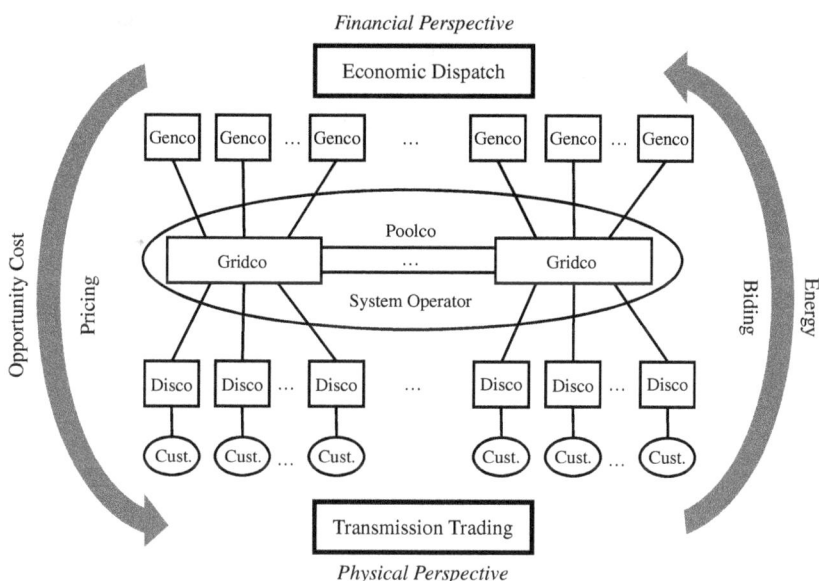

Figure 6.5. Structure of coordinated transmission trading in competitive markets (Hogan, 1998).

transactions of participants and ensures price discovery at the margin instead of on average.

The approaches are extendable to price-directed demand technologies such as vending systems with pre-paid energy accounts where individual loads make decisions on energy consumption based on pricing signals broadcasted by the system operator to all eligible loads at each market period. Combined with anticipations of demand and future pricing individual loads, even at the level of home appliances or any electricity consuming devise, can make decisions about energy use based at price level and preset thresholds for acceptable cost.

6.5 Elasticity and Price

The concept of elasticity refers to the relation between price and quantity demanded. This is generally an inverse relation, that is the lower the price the greater the quantity demanded. When it comes to electricity a similar pattern is observed. However, the functional relation between price and quantity of electricity is typically established through average costs. Utilities use their data to establish that various pricing schedules result in appropriate reduction or growth in the use of electricity. Intuitively it is reasonable to expect that consumers do not respond to marginal prices of electricity; although this is what happens with other commodities. For instance, when consumers buy apples, they respond to the actual (marginal) price listed on the apple stand, not the average price of apples in the market. Experimental observations confirm that with electricity things are different: consumers respond to average not marginal costs.

How then is it possible to have near real-time price-directed demand? In smart energy, devices make economic decisions on their own and respond to marginal not average prices. The consumer expresses preferences through price thresholds. Lifestyle needs, comfort level and economic interest are some of the variables used. For example, an air condition is programmed to use energy whose price is less than 41 cents per kilowatt-hour (the threshold), unless room temperature is greater than 30 degrees centigrade (a consumer or manufacturer suggested comfort or safety limit). The appliance is connected to the grid and the internet. The consumer has preprogrammed a set of price thresholds for electricity that can be used by the air conditioner automatically. The 24-hour day-night period is divided in a number of fixed intervals, say 48 half-hour periods. If at a given interval, the price of electricity exceeds the preprogrammed threshold of 41 cents per kilowatt-hour, the air conditioner puts itself on a standby or sleeping mode until the price falls below the threshold or a critical thermostat temperature is reached. Different price thresholds may result is a higher or lower energy bill for comparable levels of comfort.

Price-directed demand may include scenarios where a user is actually paid to use energy. For example, a plug-in hybrid automobile that has the ability to respond to pricing signals may have a negative threshold in receiving energy for battery charging. This means that the user can specify a threshold, say −3 cents per kilowatt-hour, for battery charging; if the battery is offered electricity prices above 3 cents per kwh, then it will be using the energy for charging. The negative sign indicates the user is actually paid to consume energy. For instance, during a low demand period when excess wind power has kicked in unexpectedly, and needs to be absorbed, a negative price threshold may indicate that the automobile's batteries are available to receive electricity, and the owner will be paid to make use of the energy. Again, short-run supply-demand considerations imply that cost is determined at the margin, not on average.

The capacity of devices to use marginal prices calls for the use of anticipated or forecasted prices, as well as other demand variables and elasticity.

In general, elasticity measures the responsiveness of demand to changes in price. If p denotes price and q is the quantity of electricity demanded, the elasticity of the function $p = f(q)$ with respect to q is

$$\varepsilon = \lim_{\Delta p, \Delta q \to 0} \frac{\frac{\Delta q}{q}}{\frac{\Delta p}{p}} = \left(\frac{p}{q}\right)\left(\frac{dq}{dp}\right) = \frac{p}{q}\left(\frac{1}{\frac{dp}{dq}}\right)$$

A typical demand curve is shown in Figure 6.6. When elasticity is negative, an increase in price will result in a reduction in demand. In some other cases, elasticity can be also used to measure the demand change of commodity A in response to the price change of commodity **B**. In smart energy systems, commodity A and B can be electricity at different times. For example, higher electricity price during peak hours at noon could force customers to consume more electricity in the morning or evening when prices are lower. Subsequently, an increase in price at noon will increase the demand in the morning or evening. In these cases, elasticity is positive. To differentiate these two scenarios, the former is called "self-elasticity" and the latter "cross-elasticity" (Kirschen *et al.*, 2000).

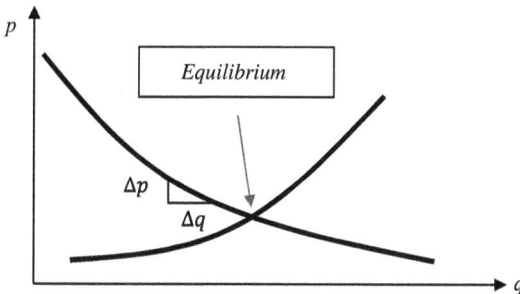

Figure 6.6. A typical demand curve, a supply curve and their intersection, that is *equilibrium*.

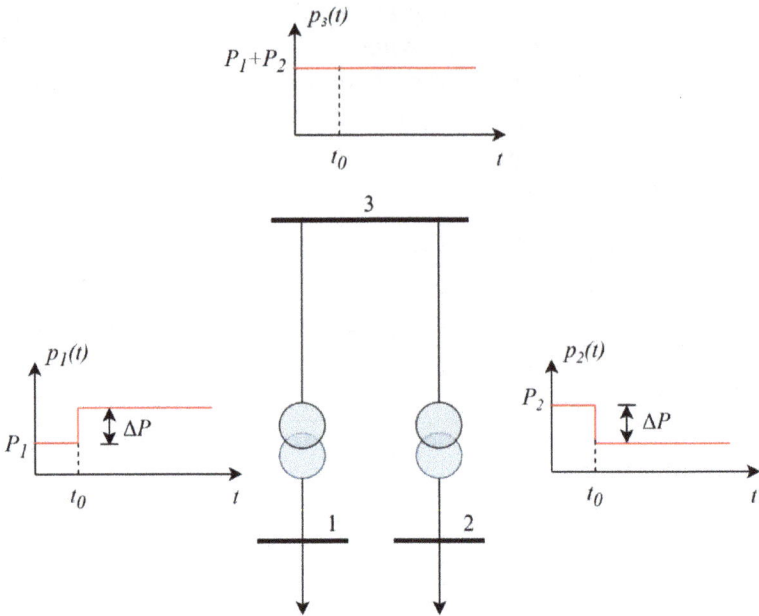

Figure 6.7. Power consumed by customers 2 and 3 completely match the power available at 3, the bus of an electrical power system 3. This is illustrative of how market-generated information (pricing signals) can be used to achieve optimal allocation of energy use.

The relation between the elasticity of demand and marginal revenue serves as a useful tool in the economics of smart energy. AI can enable the use of short-term elasticity to reshape power allocation in ways that minimize rejected power while demand closely matches the available power as shown the Figure 6.7. Pricing signals are extracted from the near instantaneous equilibrium of supply and demand. They are broadcasted to all electrical load (customers) and the pricing information can help the loads schedule their power use autonomously, thus delivering great savings to consumers while ensuring long-germ benefits to society and the environment.

Problems

Problem 6.1. Since the population cannot grow exponentially forever, a better model is the logistic which is more appropriate if the population growth is slowing down. The logistic is given by

$$P(t) = \frac{K}{1 + A e^{-rt}}$$

where, K represents the carrying capacity.

Repeat Example 6.1 using the logistic model.

Problem 6.2. If the Atomic energy transition is enabled in time, the global population growth may follow a polynomial fit. The polynomial fit is given by

$$y = a + bt + ct^2$$

Use the logistic model with the data of Example 6.1.

Problem 6.3. Consider a linear model of population growth since 1900 shown as follows. How is the value of R^2 obtained and how can it be used to compare with the exponential model of Example 6.1.

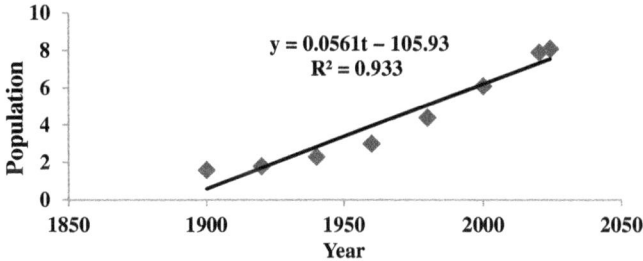

Problem 6.4. Consider the model of annual oil consumption per day (bbl/day) in the following graph. Show analytically how the value of R^2 is obtained and explain its meaning.

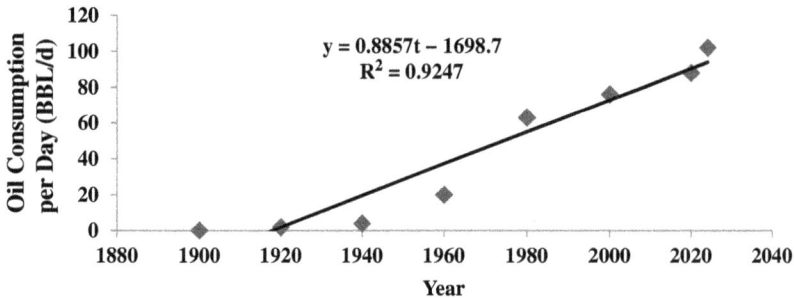

Problem 6.5. Consider a liner model of annual energy consumption per day in barrels of oil equivalent per day (boe/day), as shown in the graph below. Show analytically how R^2 is obtained and how it compares to the exponential fit of Example 6.2.

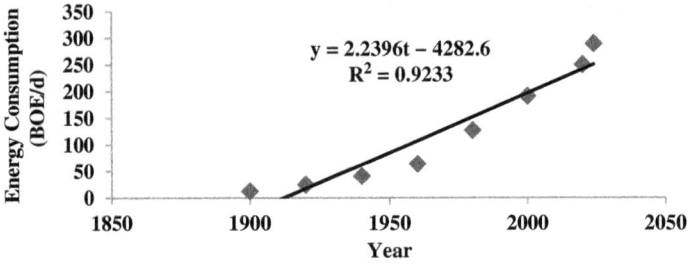

Problem 6.6. What is the correlation between an exponential model of population growth and the per capita annual oil consumption since the 1900?

Problem 6.7. What is the correlation between an exponential model of population growth and the per capita energy consumption in boe/day since the 1900?

Problem 6.8. If in 2100, all energy needs of the world were obtained from nuclear electricity, how many TWh would be needed. Assume a population plateau of 10 billion.

References

Hogan, W., *Competitive Electricity Markets: A Wholesale Primer, Report,* 1998. Available at http://www.hks.harvard.edu/fs/whogan/empr1298.pdf.

King, D. and Murray, J., "Oil's tipping point has passed," *Nature*, 481, 433–435, 2012.

Kirschen, D.S., *et al.*, "Factoring the elasticity of demand in electricity prices," *IEEE Transactions on Power Systems*, 15, 612–617, 2000.

NREL, *Financial Cases and Methods*, https://atb.nrel.gov/electricity/2024/financial_cases_&_methods, 2025.

Schumpeter, S.A., *Business Cycles: A Theoretical, Historical and Statistical Analysis of the Capitalist Process*, McGraw Hill, New York, 1964.

Index

www.ingramcontent.com/pod-product-compliance
Lightning Source LLC
Chambersburg PA
CBHW050542190326
41458CB00007B/1887